PRECALC WITH TRIGONOMETRY

Third Edition

Robert Miller

Mathematics Department
City College of New York
City University of New York

McGraw-Hill

New York   Chicago   San Francisco   Lisbon   London
Madrid   Mexico City   Milan   New Delhi
San Juan   Seoul   Singapore
Sydney   Toronto

The McGraw-Hill Companies

1 2 3 4 5 6 7 8 9 10 11 12 DOC/DOC 0 9 8 7 6 5

ISBN 0-07-145317-2

Library of Congress Cataloging-in-Publication Data

Miller, Robert, 1943–
    Precalc with trigonometry / Robert Miller. -- 3rd ed.
        p. cm. -- (Bob Miller's calc for the clueless)
    Includes index.
    ISBN 0-07-145317-2 (alk. paper)
 1. Algebra--Textbooks. 2. Trigonometry--Textbooks. 3. Functions--Textbooks. I. Title.
    QA154.3.M57 2005
    512'.13--dc22

                                        2005007304

This book is printed on recycled, acid-free paper containing a
minimum of 50% recycled de-inked fiber.

To my wife, Marlene, I dedicate this book and anything else I ever do. I love you. I love you! I LOVE YOU!!

CONTENTS

ACKNOWLEDGMENTS

I have many people to thank:

I thank my wife Marlene, who makes life worth living, who is the wind under my wings.

I thank the rest of my family: children and in-law children Sheryl and Eric, and Glenn and Wanda; grandchildren Kira, Evan, Sean; and Sarah; brother Jerry; parents and in-law parents Cele and Lee, and Edith and Siebeth.

I thank those at McGraw-Hill: Barbara Gilson, Maureen B. Walker, and Adrinda Kelly.

I thank former employees of McGraw-Hill: John Carleo, John Aliano, David Beckwith, Mary Loebig Giles, Pat Koch, Andrew Littell, and Meagan McGovern.

I thank Martin Levine of Market Source for introducing my books to McGraw-Hill.

I thank Daryl Davis, Bernice Rothstein, Sy Solomon, and Dr. Robert Urbanski.

As usual the last thanks go to three terrific people: a great friend Gary Pitkofsky, another terrific friend and fellow teacher David Schwinger, and my sharer of dreams, my cousin Keith Robin Ellis.

TO THE STUDENT

Congratulations for reaching precalculus and trig!!! Most students in the world never reach this level.

This book was written for you—not your teacher, not your next-door neighbor, not for anyone but you.

I have tried to make the examples and explanations as clear as I can. However, as much as I don't like to admit it, I'm not perfect. So if you have questions, visit my Web site at

www.mathclueless.com

You may also e-mail me at

bobmiller@mathclueless.com

There is a problem you might have with precalculus. Many students' algebraic background is so poor that they have trouble with precalculus. (I hope you are not one of them.) A trip to *Algebra for the Clueless* may be necessary.

Now enjoy the book, and learn!

Bob Miller

CHAPTER 1

LINES, THE STRAIGHT KIND

In elementary algebra, you learned how to graph points and how to graph lines. We will briefly review graphing and then attack the harder problem, finding the equation of a straight line.

STANDARD FORM

When we look at an equation, we should instantly know that it is a straight line. Any equation in the form $Ax + By = C$, where A, B, and C are numbers and A and B are not both equal to 0, is a straight line.

NOTE

In some books, standard form is $Ax + By + C = 0$, so standard form isn't standard. Isn't that funny?!

Let's make sure we understand this by giving some examples that are straight lines and some that are not.

EXAMPLE 1—

A. $3x - 4y = 7$

B. $5x = 9$

C. $x/3 - y/7 = 7$

D. $3/x + 5/y = 9$

E. $xy = 7$

F. $x^2 - 3y = 5$

A, B, and C are lines. Coefficients may be negative or fractions or, as in B, B = 0. D is not a straight line since the letters are in the bottom. E is not a straight line since the variables are multiplied. F is not a straight line since the exponent of each letter must be 1.

DEFINITION

Slope: The slope (or slant) of a line is defined by $m = (y_2 - y_1)/(x_2 - x_1)$.

NOTES

1. The letter m is always used for the slope.

2. The 1s and 2s are subscripts, standing for point 1 and point 2. The 1 and the 2 mean x_1, x_2, y_1, and y_2 and stand for numbers, not variables, but I will not tell you what they are yet.

3. The calculus notation is $m = \Delta y/\Delta x$, where Δ is the Greek letter delta and means the change in y over the change in x.

4. The y's are always on top.

EXAMPLE 2 ON THE SLOPE—

We will graph the line that joins the points after this example and discuss it further. Find the slope between

A. (2, 3) and (6, 12)

B. (4, −3) and (−1, 3)

C. (1, 3) and (6, 3)

D. (2, 5) and (2, 8)

SOLUTIONS—

A. (2, 3) and (6, 12)

$$\underset{x_1\ \ y_1}{\uparrow\ \uparrow} \qquad \underset{x_2\ \ y_2}{\uparrow\ \uparrow}$$

$$m = \frac{y_2 - y_1}{x_2 - x_1} = \frac{12 - 3}{6 - 2} = \frac{9}{4}$$

B. (4, −3) and (−1, 3)

$$\underset{x_1\ \ y_1}{\uparrow\ \uparrow} \qquad \underset{x_2\ \ y_2}{\uparrow\ \uparrow}$$

$$m = \frac{y_2 - y_1}{x_2 - x_1} = \frac{3 - (-3)}{-1 - 4} = \frac{6}{-5}$$

NOTE

In 3 − (−3), the first minus sign is the one in the equation, and the second comes from the fact that y_1 is negative. *Be careful!*

C. (1, 3) and (6, 3)   m = (3 − 3)/6 − 1 = 0/5 = 0

D. (2, 5) and (2, 8)   m = (8 − 5)/(2 − 2) = 3/0— undefined, no slope, or infinite slope

THE NON-MATH EXPLANATION FOR EXAMPLE 2—

A. If you walk in the direction of the arrow, when you get to the line and have to walk up, you always have a *positive slope.*

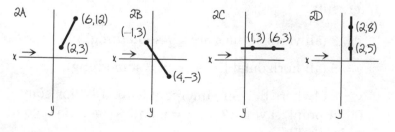

> **NOTE**
>
> *Until you get good at this, label the points just like I did. Also, it does not matter which is point 1 and which is point 2.*

B. If you have to walk down, you have a *negative slope.*

C. Horizontal lines have m = 0.

D. Vertical lines have no slope or infinite slope.

DEFINITION

Intercepts: The *x intercept,* where the line hits the x axis, is where y = 0. The *y intercept* is where x = 0.

EXAMPLE 3—

Given 3x − 4y = 12. Find the intercepts and graph the line.

x intercept: y = 0. 3x = 12. x = 4. The point is (4, 0). (x coordinate is always first.)

y intercept: x = 0. −4y = 12. y = −3. The point is (0, −3), and the graph is as shown:

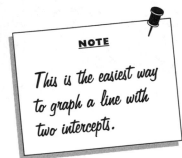

NOTE

This is the easiest way to graph a line with two intercepts.

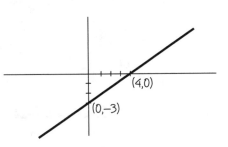

There are three exceptions, the ones with only one intercept:

x = 3 (all vertical lines are x = something).

y = 5 (all horizontal lines are y = something).

y = 3x [where the only intercept is (0, 0)]. Pick some other point, say, x = 2. So y = 6 and we get the second point (2, 6).

Graphs are all included in the picture here:

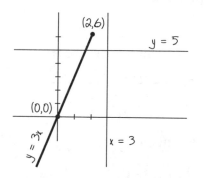

We are almost ready to find the equation of a line. There are four basic forms of a line. The two main ones are the point-slope and the slope-intercept.

DEFINITION

Slope-intercept: $y = mx + b$. When you solve for y, the coefficient of x is the slope. b is the y intercept—when $x = 0$, $y = b$. This is the most common form in high school books.

DEFINITION

Point-slope: $m = (y - y_1)/(x - x_1)$. You are given the slope and a point (x_1, y_1). The point-slope is given in different forms in different books. I will try to convince you that this form is the best to use because it eliminates many of the arithmetic fractions that tend to bother too many of you.

We will now do some problems of this type. Although they are not long problems, most students have difficulty at first. Don't get discouraged if you do.

EXAMPLE 4—

Find the equation of the line through the points (2, 3) and (7, 11) and write the answer in standard form.

We will do this problem two different ways. I think you may be convinced the point-slope method is better.

Point-slope **(2, 3) and (7, 11)**	**Slope-intercept** **(2, 3) and (7, 11)**
$m = \dfrac{(11-3)}{(7-2)} = \dfrac{8}{5}$	$m = \dfrac{(11-3)}{(7-2)} = \dfrac{8}{5}$
$m = \dfrac{y - y_1}{x - x_1}$	$y = mx + b$
$\dfrac{8}{5} = \dfrac{y-3}{x-2}$	$y = \left(\dfrac{8}{5}\right)x + b$
$8(x-2) = 5(y-3)$	$3 = \left(\dfrac{8}{5}\right)(2) + b$
$8x - 16 = 5y - 15$	$3 = \left(\dfrac{16}{5}\right) + b$
$8x - 5y = 1$	
	$b = 3 - \dfrac{16}{5} = \dfrac{15}{5} - \dfrac{16}{5} = -\dfrac{1}{5}$
	$y = \left(\dfrac{8}{5}\right)x - \dfrac{1}{5}$
	$5y = 8x - 1$
	$-8x + 5y = -1 \text{ or } 8x - 5y = 1$

The reason the point-slope method is easier for most is that the only fractional skill needed is cross multiplication,* a skill most math students have. The slope-intercept method requires a number of fractional skills. About 90% of the time, point-slope is better.

At this stage of your mathematics, all of you should put the lines in standard form to practice your algebra

---

*See the third to the fourth lines of the point-slope column.

skills. However, to save space and keep the book friendlier, I will leave the answer in either of these two forms.

NOTE

Most of this chapter is on the 21st century MATH SAT.

EXAMPLE 5—

Write the equation of the line with slope 3, y intercept 7.

$y = mx + b$ is the easiest. $y = 3x + 7$.

EXAMPLE 6—

Write the equation of the line with slope −4 and x intercept 9.

Tricky. $m = -4$ and point (9, 0). Use point-slope:

$$m = \frac{y - y_1}{x - x_1} \qquad -4 = \frac{y - 0}{x - 9}$$

EXAMPLE 7, PART A—

Find the equation of the line parallel to $3x + 4y = 5$ through (6, 7).

EXAMPLE 7, PART B—

Find the equation of the line perpendicular to $3x + 4y = 5$ through (8, −9).

SOLUTIONS A AND B—

Parallel lines have the same slope. Perpendicular lines have negative reciprocal slopes. In either case we must solve for y.

$3x + 4y = 5$. $4y = -3x + 5$. $y = (-3/4)x + 5/4$. We only care about the coefficient of x, which is the slope. The slope is −3/4.

A. *Parallel* means equal slopes. m = –3/4; the point is (6, 7). The equation is

$$\frac{-3}{4} = \frac{y - 7}{x - 6}$$

B. *Perpendicular* means the slope is the negative reciprocal. m = +4/3; the point is (8, –9). The equation is

$$\frac{4}{3} = \frac{y - (-9)}{x - 8} \qquad \text{or} \qquad \frac{4}{3} = \frac{y + 9}{x - 8}$$

CHAPTER 2

QUADRATIC EQUATIONS

A topic you should have either partially or totally covered before is how to solve quadratic formulas, but I can't leave this to chance. Let's do it.

SOLVING A QUADRATIC (OR HIGHER) BY FACTORING

NOTE

This topic is on the 21st century MATH SAT.

If the product of two numbers is 0, at least one of the numbers must be 0. In letters, if (a)(b) = 0, either a = 0 or b = 0.

EXAMPLE 1—

(x − 4)(x + 5) = 0

x − 4 = 0, so x = 4. x + 5 = 0, so x = −5. The two answers are 4 and −5.

EXAMPLE 2—

x(x − 4)(x + 7)(2x − 9)(3x + 1) = 0

Setting each factor equal to 0, the answers are x = 0, 4, −7, 9/2, and −1/3.

Practice doing this in your head!!!!! It's not hard if you try.

EXAMPLE 3—

$x^2 - 2x = 8$

SOLUTION—

Get everything on one side—arrange terms highest exponent to lowest exponent.

$x^2 - 2x - 8 = 0$

$(x - 4)(x + 2) = 0$

So x = 4 and −2.

Now factor and set each equal to 0 as before.

EXAMPLE 4—

A cubic—three solutions.

$x^3 - 7x^2 - 8x = 0$

SOLUTION—

$x(x - 8)(x + 1) = 0$

So x = 0, 8, −1.

EXAMPLE 5—

At first this looks rather tame. However, the factored form must equal 0. So there is a bit of work to do.

$(x - 2)(x - 3) = 2$

SOLUTION—

Multiply out, move everything to one side, and then refactor.

$x^2 - 5x + 6 = 2$

$x^2 - 5x + 4 = 0$

$(x - 4)(x - 1) = 0$

The answers are x = 4 and 1.

If you are not terrific at factoring, you should *practice, practice, practice!!!!!!*

THE QUADRATIC FORMULA

What if the equation does not factor? If it is a quadratic, there is a formula that will work. We will derive it here because you should see it and because some teachers actually make students do this (many times) by the method used to prove this formula. We need a brief introduction.

Get a perfect square: $(x + k)^2 = x^2 + 2kx + k^2$. If the coefficient of x^2 is 1, take half of the coefficient of x (half of 2k, which is k) and square it to get k^2, the sum of which factors into two identical terms: $x^2 + 10x$. Half of 10 is 5. 5^2 is 25. $x^2 + 10x + 25 = (x + 5)(x + 5) = (x + 5)^2$.

How about a little more? Suppose we had $x^2 + ax$. Half of a is (1/2)a. Square it—$1/4a^2$. We have $x^2 + ax + (1/4)a^2 = (x + 1/2a)^2$! Now let's get back to our problem.

We will solve the quadratics $3x^2 - 7x - 6 = 0$ and $ax^2 + bx + c = 0$ simultaneously, preceded on the right by what we are doing on the next step.

$3x^2 - 7x - 6 = 0$	$ax^2 + bx + c = 0$	**Divide by the coefficient of x^2.**
$\dfrac{3x^2}{3} - \dfrac{7x}{3} - \dfrac{6}{3} = \dfrac{0}{3}$	$\dfrac{ax^2}{a} + \dfrac{bx}{a} + \dfrac{c}{a} = \dfrac{0}{a}$	**Get the term without x to the other side.**
$x^2 - \dfrac{7x}{3} = 2$	$x^2 + \dfrac{bx}{a} = \dfrac{-c}{a}$	**Complete the square. Take half the coefficient of x, square it, and add to both sides.**
$x^2 - \dfrac{7x}{3} + \left(-\dfrac{7}{6}\right)^2 = \left(-\dfrac{7}{6}\right)^2 + 2$	$x^2 + \dfrac{bx}{a} + \left(\dfrac{b}{2a}\right)^2 = \left(\dfrac{b}{2a}\right)^2 - \dfrac{c}{a}$	**Factor the left side and do arithmetic and algebra on the right side.**
$\left(x - \dfrac{7}{6}\right)^2 = \dfrac{49}{36} + \dfrac{72}{36} = \dfrac{121}{36}$		
	$\left(x + \dfrac{b}{2a}\right)^2 = \dfrac{b^2}{4a^2} - \dfrac{c(4a)}{a(4a)}$	**Take the square root of both sides.**
	$= \dfrac{b^2 - 4ac}{4a^2}$	

Isolate x and get the two solutions (roots).

$$x - \frac{7}{6} = \pm \frac{11}{6}$$

$$x = \frac{7}{6} \pm \frac{11}{6}$$

$$x_1 = \frac{7}{6} + \frac{11}{6} = \frac{18}{6} = 3$$

$$x_2 = \frac{7}{6} - \frac{11}{6} = -\frac{4}{6} = \frac{-2}{3}$$

$$x = \frac{b}{2a} \pm \frac{\sqrt{b^2 - 4ac}}{2a}$$

$$x = \frac{-b}{2a} \pm \frac{\sqrt{b^2 - 4ac}}{2a}$$

$$x_1 = \frac{-b + \sqrt{b^2 - 4ac}}{2a}$$

$$x_2 = \frac{-b - \sqrt{b^2 - 4ac}}{2a}$$

The theorem we just proved, the quadratic formula, states that given the equation $ax^2 + bx + c = 0$, with $a \neq 0$, the roots will be

$$x = \frac{-b \pm \sqrt{b^2 - 4ac}}{2a}$$

where a is the coefficient of the x^2 term, b is the coefficient of the x term, and c is the term without x.

Let's do some examples. The first one is the above example using the quadratic formula.

EXAMPLE 6—

Solve $3x^2 - 7x - 6 = 0$ using the quadratic formula.
$a = 3, b = -7, c = -6$.

SOLUTION—

$$x = \frac{-(-7) \pm \sqrt{(-7)^2 - 4(3)(-6)}}{2(3)}$$

$$x = \frac{7 \pm \sqrt{121}}{6}$$

As before, $x = 3$ and $-2/3$.

NOTE 1

After we prove this formula, look how easy it is to use.

NOTE 2

This problem can be factored. You should always try to factor first. $3x^2 - 7x - 6 = 0$. $(3x + 2)(x - 3) = 0$. Again, $x = 3, -2/3$. The main reason to use the formula is if the quadratic does *not* factor.

EXAMPLE 7—

$2x^2 + 5x + 1 = 0$. $a = 2, b = 5, c = 1$.

SOLUTION—

Sooooo . . .

$$x = \frac{-5 \pm \sqrt{5^2 - 4(2)(1)}}{2(2)}$$

$$x = \frac{-5 \pm \sqrt{17}}{4}$$

This is the major use for this formula.

EXAMPLE 8—

$3x^2 + 5x + 7 = 0$. $a = 3, b = 5, c = 7$.

SOLUTION—

$$x = \frac{-5 \pm \sqrt{5^2 - 4(3)(7)}}{2(3)} = \frac{-5 \pm \sqrt{-59}}{2(3)} = \frac{-5 \pm i\sqrt{59}}{6}$$

NOTE

$$\sqrt{-1} = i; \quad \sqrt{-59} = \sqrt{-1}\sqrt{59} = i\sqrt{59}$$

This is part of a topic called *imaginary (complex) numbers*. This is discussed in most elementary algebra books. Basic imaginary-number properties are very easy (all of algebra should be that easy).

Most of you will do very well with this formula. However, very few students know it as well as I want my students to know it. Try this little exercise to see if you *really* know what the formula says.

For each of the following, tell what a is, what b is, and what c is. After you do one problem, look at the answers before you do the next.

PROBLEMS

1. $3x^2 - x - 7 = 0$. a = ?, b = ?, c = ?

2. $7x - x^2 = 0$. a = ?, b = ?, c = ?

3. $a^2x^2 - bx + c = 0$. a = ?, b = ?, c = ?

4. (Really tough) $-x^2 + b^2 + 7 = 0$. a = ?, b = ?, c = ?

5. $3x^2 + bx + 7x - 5 = 0$. a = ?, b = ?, c = ?

ANSWERS

1. $a = 3, b = -1, c = -7$.

2. $a = -1, b = 7, c = 0$.

3. $a = a^2, b = -b, c = c$.

4. $a = -1, b = 0, c = b^2 + 7$!!

5. $a = 3, b = (b + 7), c = -5$.

If you understand the last five examples, you really understand the formula.

CHAPTER 3

INEQUALITIES, LINEAR AND QUADRATIC

LINEAR INEQUALITIES

Let's review linear inequalities. This is a topic I hope you are familiar with, but just to make sure, we will go over it.

Linear inequalities are done exactly the same way as linear equalities except that when you multiply or divide by a negative number, the order is reversed.

NOTE

$8 < 12$; then $8/2 < 12/2$ since $4 < 6$, but $8/-2 > 12/-2$ since $-4 > -6$.

EXAMPLE 1—

$4(3x - 5) - 2(4x - 6) \leq 6x - 14$

$12x - 20 - 8x + 12 \leq 6x - 14$

$4x - 8 \leq 6x - 14$

$-2x - 8 \leq -14$

$-2x \leq -6$

$x \geq +3$

Multiply out parentheses. On each side, combine like terms.

Add $-6x$ to each side.

Add $+8$ to each side.

Divide each side by -2; the order reverses.

NOTE

These topics are on the 21st century MATH SAT.

The graph looks like this:

The solid dot means that 3 is part of the answer. An open dot would mean that it was not.

EXAMPLE 2—

We are solving a double inequality. The x must be left in the middle. Add 7 to each term.

$$-3 < 2x - 7 \le 16$$

Then divide each term by 2.

$$4 < 2x \le 23$$

$$2 < x \le \frac{23}{2}$$

QUADRATIC INEQUALITIES

This is one of my favorite topics. I think I teach it about as well as anyone. Most books do a rather poor job. I think you'll like this topic after these few pages.

EXAMPLE 3—

Get everything to one side, coefficient of x^2 positive, and then factor (in some examples you might need the quadratic formula).

$$x^2 - 7x > 8$$

$$x^2 - 7x - 8 > 0$$

$$(x - 8)(x + 1) > 0$$

There are three regions on the graph—$x > 8$, $-1 < x < 8$, and $x < -1$. The technique is to substitute a number in each of the regions; but *do not do the arithmetic* since we only care about the *sign* of the answer.

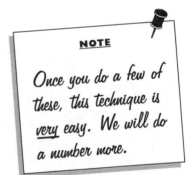

Draw a graph with an open dot on 8 and an open dot on -1. We are solving the equality $(x - 8)(x + 1) = 0$, so $x = 8, -1$ doesn't solve the given inequality.

1. Take a number bigger than 8—say, 10. Substitute in $(x - 8)(x + 1) > 0$. $10 - 8$ is positive, $10 + 1$ is positive, and the product is positive, that is, greater than 0, so the region $x > 8$ *is* part of the answer.

2. Take a number between -1 and 8—say, 3. $3 - 8$ is negative and $3 + 1$ is positive, and the product is negative, that is, *not* greater than 0. $-1 < x < 8$ is not part of the answer.

3. Take a number less than -1—say, -100. $-100 - 8$ is negative and $-100 + 1$ is negative, and the product is positive, that is, greater than 0, so $x < -1$ is part of the answer.

The graph looks like this:

NOTE

Once you do a few of these, this technique is very easy. We will do a number more.

$x < -1$ or $x > 8$

EXAMPLE 4—

$(x - 2)(x - 4)(x - 6)(x - 8) \leq 0$

With other techniques this problem would take a very long time, but after we finish this example, this will take you 10 seconds.

SOLUTION—

Solid dot on 2, 4, 6, and 8, since the product can equal 0.

Substitute x = 9. (9 – 2)(9 – 4)(9 – 6)(9 – 8) is the product of all positive terms, so the product couldn't be negative or 0, so x > 8 is not part of the answer.

x = 7. All terms are positive except 7 – 8. The whole product is negative. $6 \le x \le 8$ is part of the answer.

x = 5. Two negative terms: (5 – 8)(7 – 8). The product is positive. $4 < x < 6$ is not part of the answer.

x = 3. Three negative terms. $2 \le x \le 4$ is part of the answer.

x = 1. Four negative terms. $x < 2$ is not part of the answer.

So the total answer is $2 \le x \le 4$ or $6 \le x \le 8$ and the graph is

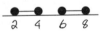

The next problem is even easier. If all the exponents are 1 (or odd numbers), every other region is part of the answer. So if x > 8 is no, then $6 \le x \le 8$ must be yes, $4 < x < 6$ no, $2 \le x \le 4$ yes, and x < 2 no!!!!!!! Easy, huh?!!!!!!

EXAMPLE 5—

$$\frac{(x-1)(x-3)(x-5)}{(x-7)(x-9)} \ge 0$$

Look how easy. Solid dot on 1, 3, and 5 since the top of the fraction can equal 0. Open dot on 7 and 9 since the bottom of a fraction can never be 0.

All we need is to substitute one number—x = 10. Since everything is positive, the product is positive and x > 9 is part of the answer. Since all the exponents are odd, every other region is part of the answer. $7 \le x \le 9$

is not, $5 \leq x < 7$ is, $3 < x < 5$ is not, $1 \leq x \leq 3$ is, and $x < 1$ is not.

The solution is $x > 9$ or $5 \leq x < 7$ or $1 \leq x \leq 3$. The graph is

NOTE

In doing these problems, graph the points, here 1, 3, 5, 7, and 9, with appropriate open or closed dots first, determine the regions next, and write the algebraic solutions last.

EXAMPLE 6—

$$\frac{2x - 5}{x - 3} > 1$$

$$\frac{2x - 5}{x - 3} - 1 > 0 \quad \text{or} \quad \frac{2x - 5}{x - 3} - \frac{x - 3}{x - 3} = \frac{x - 2}{x - 3} > 0$$

Substituting $x = 5$ and noting that it *is* part of the answer and that all the exponents are 1 so every other region is the answer, the algebraic answer is $x > 3$ or $x < 2$ and its graph is

EXAMPLE 7—

$$\frac{2x - 6}{x - 3} \leq \frac{x - 2}{x - 3}$$

$$\frac{2x - 6}{x - 3} - \frac{x - 2}{x - 3} = \frac{x - 4}{x - 3} \leq 0$$

Since both bottoms are the same, get everything to the side where the coefficient of the top is going to be positive, and subtract the fractions.

Solid dot on 4 since the top can equal 0.

Open dot on 3 since the bottom can never be 0.

We proceed as before: solid dot on 4 and open dot on 3. Substitute a number bigger than the rightmost point (4)—say, $x = 6$. We get $(6 - 4)/(6 - 3)$, which is certainly not less than or equal to 0. So the right end is not part of the answer. Since all the exponents are odd, every other region is part of the answer. In this case only the middle is the answer: $3 < x \le 4$. Its graph is

EXAMPLE 8—

Only the algebra is messier. Getting all the terms to one side, you should subtract the fractions by sight.

Note:

$$\frac{a}{b} - \frac{c}{d} = \frac{ad - bc}{bd}$$

With a little practice you can do this easily!!!

$$\frac{x - 3}{x - 1} > \frac{x - 5}{x - 2}$$

$$\frac{x - 3}{x - 1} > \frac{x - 5}{x - 2} \qquad \text{or}$$

$$\frac{x - 3}{x - 1} - \frac{x - 5}{x - 2} = \frac{(x - 3)(x - 2) - (x - 5)(x - 1)}{(x - 1)(x - 2)}$$

$$= \frac{(x + 1)}{(x - 1)(x - 2)} > 0$$

We proceed as before: open dot on -1, 1, and 2. Substitute a number to the right of 2—say, $x = 5$. Everything is positive, so $x > 2$ is part of the answer. Again, all exponents are odd, so every other region is part of the answer. The answer is $x > 2$ or $-1 < x < 1$, and its graph is

EXAMPLE 9—

$$\frac{(x^2 + 15)(x - 2)^3(x - 4)^4(x - 6)^5}{(x - 8)^6} \geq 0$$

Now don't panic. This is a *lot* easier than it looks, especially if you did the previous examples. First note that no matter what x is, $x^2 + 15$ is always positive. In an inequality, this means you can discount it. So throw it out!! The problem is now

$$\frac{(x - 2)^3(x - 4)^4(x - 6)^5}{(x - 8)^6} \geq 0$$

Not much better yet. However, it really is easy. Solid dot on 2, 4, and 6 and open dot on 8 as before.

Substitute x = 9. All terms are positive—x > 8 is part of the answer.

x = 7. All terms are positive—$6 \leq x < 8$ is also part of the answer.

x = 5. One term is negative $(5 - 6)^5$, so the expression is negative—$4 < x < 6$ is not part of the answer.

x = 3. The same one term is negative, so $2 < x < 4$ is not part of the answer.

x = 1. Two terms are negative, and the product is positive—$x \leq 2$ is part of the answer.

The answer is $x \leq 2$ or $x = 4$ or $6 \leq x < 8$ or $x > 8$. Its graph looks like this:

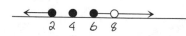

NOTE

There are several weird things that happen here. x > 8 is part of the answer and $6 \leq x < 8$ is part of the answer, but x = 8 is not part of the answer; x = 4 is part of the answer, but the region to the left and the region to the right are not part of the answer. We have an isolated point. Lots of fun things are possible when exponents are both even and odd. But with a little practice, all these problems should be easy.

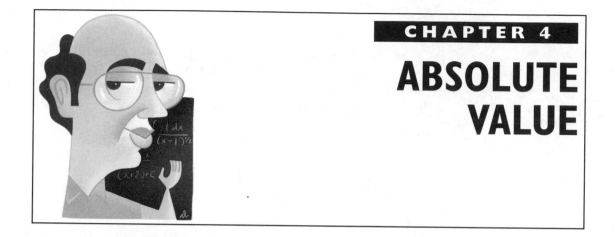

ABSOLUTE VALUE

There are two definitions of *absolute value*. We will give both since both are used.

DEFINITIONS

Absolute value definition 1:

$|x| = x$ if $x > 0$

$\quad = -x$ if $x < 0$

$\quad = 0$ if $x = 0$

Using numbers, $|9| = 9$ since $9 > 0$; $|-3| = -(-3) = 3$ since $-3 < 0$; and $|0| = 0$.

Absolute value definition 2:

$|x| = \sqrt{x^2}$

that is,

$|-6| = \sqrt{(-6)^2} = \sqrt{36} = 6$

You might say the second definition is much easier, and you would be correct. However, the first definition

NOTE

This entire chapter is on the 21st century MATH SAT.

23

is almost always easier to use, as we will see, and in some cases the second definition is virtually unusable.

Let us list some properties of absolute value. They all can be verified by substituting numbers.

Property 1: $|ab| = |a| \, |b|$.

Property 2: $\left|\dfrac{a}{b}\right| = \dfrac{|a|}{|b|}$, if $b \neq 0$.

Property 3: $|a - b| = |b - a|$.

Property 4: The triangle inequality $|a + b| \leq |a| + |b|$.

The last property is most interesting and is used later in your math. If a is 0, or if a and b are both positive, or if a and b are both negative, the statement is an equality. However, if, let us say, $a = 6$ and $b = -2$, it is an inequality. You might ask, "Why the heck do I want an inequality?" There are two reasons: (1) that is all you can get; (2) if we can show $x \leq y$ *and* $y \leq x$, we get $x = y$! Pretty sneaky, eh? However, this is for later. But it is important to preview it now so that later this will not be scary.

We are now ready to do the absolute value problems.

EXAMPLE 1—

$|x - 4| = 6$

Using definition 1, $x - 4 = 6$ or $x - 4 = -6$ since the absolute value of 6 and -6 is 6. The answers are $x = 10$ and $x = -2$.

Let's show how messy it is to use the other definition:

$(x - 4)^2 = 6^2$

or $(x - 4)^2 = 36$

or $x^2 - 8x + 16 = 36$

or $x^2 - 8x - 20 = 0$

Factoring, we get $(x - 10)(x + 2) = 0$. Again the answers are 10 and –2 . . . except we have done much more work!!!!

EXAMPLE 2—

$|3x - 7| = 0$

There is only one solution: $3x - 7 = 0$ or $x = 7/3$.

EXAMPLE 3—

$|4x + 7| = -5$

No solution since the absolute value is never negative.

EXAMPLE 4—

$|x - 8| < 3$

To give you an idea of what this means, we will use a picture. The absolute value can stand for the distance between two points. The example says we are looking for all points x that differ from 8 by less than 3. Here is the picture:

NOTE

"Outsides" are greater than picture. Algebraically, $-3 < x - 8 < 3$ or $5 < x < 11$.

EXAMPLE 5—

$|8x - 4| < -9$

No solution since the absolute value is never negative.

EXAMPLE 6—

$|5 - 4x| \geq 6$

Since I don't like the way it looks, property 3 allows me to write it as $|4x - 5| \geq 6$. There are two parts: $4x - 5 \geq 6$ or $4x - 5 \leq -6$. The solution is $x \geq 11/4$ or $x \leq -1/4$. Its graph looks like this:

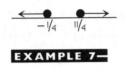

EXAMPLE 7—

$|2x - 5| \geq -3$

The solution is all real numbers since the absolute value is always bigger than a negative number.

CHAPTER 5

EXPONENTS— NEGATIVE, FRACTIONAL

One of the easiest yet most forgettable topics concerns exponents. We will try to do all the new stuff and at the same time go over problems from the past.

DEFINITION

Negative exponents: $a^{-n} = 1/a^n$, $b^{-3} = 1/b^3$, and $1/m^{-4} = m^4$. In other words, *negative exponent* means reciprocal and has nothing to do with negative numbers!

EXAMPLES I AND 2—

$5^{-2} = 1/5^2 = 1/25$

$(-4)^{-3} = 1/(-4)^3 = 1/-64 = -1/64$

DEFINITION

Fractional exponents: $a^{p/r}$, where p is the power and r is the root.

Always do the root first!!!

EXAMPLE 3—

$27^{2/3} = (\sqrt[3]{27})^2 = 3^2 = 9$

EXAMPLE 4—

$16^{-(3/2)} = 1/16^{3/2} = 1/(\sqrt{16})^3 = 1/4^3 = 1/64$

RULE 1

$a^m a^n = a^{m+n}$. If the bases are the same, when you multiply you add the exponents and leave the base unchanged.

EXAMPLE 5—

$(5a^{2/3})(4a^{3/2}) = 20a^{2/3 + 3/2} = 20a^{13/6}$

Remember—coefficients are multiplied; exponents are added.

RULE 2

$(a^m)^n = a^{mn}$. Power to a power, you multiply exponents.

EXAMPLE 6—

$(4a^{5/7})^{3/2} = 4^{3/2}a^{15/14} = 8a^{15/14}$

Don't forget the numerical coefficient!

RULE 3

$a^m/a^n = a^{m-n}$ or $1/a^{n-m}$

If the bases are the same, *divide* means subtract the exponents, leaving the base the same.

EXAMPLE 7—

$$\frac{a^6 b^{-3} c^{-7}}{a^{-9} b^4 c^{-5}} = \frac{a^6 a^9 c^5}{b^3 b^4 c^7} = \frac{a^{15}}{b^7 c^2}$$

RULE 4

$(ab)^n = a^n b^n$

NOTE

If the problem contains only multiplication and division, negative exponents in the top become positive exponents in the bottom and negative exponents in the bottom become positive exponents in the top.

EXAMPLE 8—

$(a^4b^{-3})^{-2} = a^{-8}b^6 = b^6/a^8$

RULE 5

$(a/b)^n = a^n/b^n$

EXAMPLE 9—

$(3a^4/b^{-5})^{-3} = 3^{-3}a^{-12}/b^{15} = 1/27a^{12}b^{15}$

EXAMPLE 10—

$$(a^4b^{-9}/b^{-5}a^{-3})^{-2} = \left(\frac{a^4b^5a^3}{b^9}\right)^{-2} = \left(\frac{a^7}{b^4}\right)^{-2}$$

$$= \left(\frac{b^4}{a^7}\right)^2 = \frac{b^8}{a^{14}}$$

EXAMPLE 11—

$$\frac{a^{-2} - b^{-2}}{a^{-1} - b^{-1}}$$

The problem is harder because of the subtraction. It is also a favorite of mine since it contains a number of skills in a relatively short problem.

$$\frac{a^{-2} - b^{-2}}{a^{-1} - b^{-1}} = \frac{\dfrac{1}{a^2} - \dfrac{1}{b^2}}{\dfrac{1}{a} - \dfrac{1}{b}}$$

Definition of negative exponent.

$$= \frac{\dfrac{1}{a^2}\dfrac{a^2b^2}{1} - \dfrac{1}{b^2}\dfrac{a^2b^2}{1}}{\dfrac{1}{a}\dfrac{a^2b^2}{1} - \dfrac{1}{b}\dfrac{a^2b^2}{1}}$$

This is a small review of complex fractions. Find the *least common denominator* (LCD) of all the fractions, and multiply each fraction top and bottom by the LCD. Multiply and simplify each fraction.

$$= \frac{b^2 - a^2}{ab^2 - ba^a}$$

$$= \frac{(b - a)(b + a)}{ab(b - a)} = \frac{b + a}{ab}$$

Factor and cancel.

At this point I usually give my classes two short, non-counting quizzes to see if they really know the laws. You might try them.

Quiz I

1. $(4a^3)(3a^4)$

2. $4a^3 + 3a^4$

3. $(4a^3)(4a^3)$

4. $4a^3 + 4a^3$

5. $(4a^3)^3$

The purpose of the quiz is to see whether you know the difference between the adding and multiplying rules.

ANSWERS

1. Multiplying. Multiply coefficients, add exponents. The answer is $12a^7$.

2. Adding. We can only add like terms—these are unlike. The answer is $4a^3 + 3a^4$.

3. Multiplying. $16a^6$.

4. Adding. Like terms—add coefficients, *leave exponents alone*. $8a^3$.

5. Rule 2. $64a^9$.

The second quiz is even harder. Try it.

Quiz 2

1. $4^7 4^{11}$

2. $b^{2x + 3} b^{4x + 7}$

3. $(a^{4x+3})(a^{7x})/a^c$ (only one a in the answer)

4. $2^n + 2^n$

5. $3^n + 3^n$

6. $3^n + 3^n + 3^n$

ANSWERS

1. Multiplying. The exponents are added, but the base stays the same. 4^{18}.

2. Same as 1. b^{6x+10}.

3. Same as 1 plus if you divide, you subtract exponents. $a^{11x+3-c}$.

4. Very tough adding problem. When you add, you add the coefficients, leaving the base alone—the coefficient is 1!!! Second part: base the same (multiply), add exponents. $2^n + 2^n = (1)2^n + (1)2^n = 2(2^n) = 2^1 2^n = 2^{n+1}$.

5. $3^n + 3^n = 1(3^n) + 1(3^n) = 2(3^n)$ since the bases are different, but . . .

6. $3^n + 3^n + 3^n = 3(3^n) = 3^1 3^n = 3^{n+1}$!!!!!

If you know all of these, then you really know your exponents!!!!!!!!

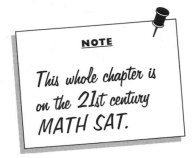

NOTE

This whole chapter is on the 21st century MATH SAT.

CHAPTER 6

GEOMETRIC FORMULAS AND FACTS YOU MUST KNOW

This chapter contains geometric formulas and facts that I hope you know very, very well. Some are needed for precalc, some are needed for trig, and all are needed for calculus.

SQUARES, RECTANGLES, BOXES, AND CUBES

Area A = bh (base times height). *Perimeter* p = 2b + 2h. (In case you forget the formula, remember that *perimeter* means to add up all the sides.) *Diagonal* d = $\sqrt{b^2 + h^2}$.

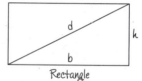

A = s^2 (side squared). p = 4s. d = $s\sqrt{2}$.

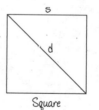

Volume V = ℓwh (length times width times height). d = $(\ell^2 + w^2 + h^2)^{1/2}$.

  Surface area SA = 2ℓw   + 2wh + 2ℓh.
                    top, bottom   ends   back, front

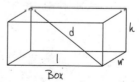

V = e^3 (edge cubed). SA = $6e^2$. d = $e\sqrt{3}$.

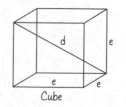

Rectangle

Square

Box

Cube

CIRCLES

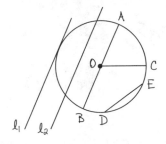

O = center of the circle, $\overline{OC}$ = *radius* r. $\overline{AB}$ = *diameter* d. r = (1/2)d. A = πr^2 (π is pronounced pi). c = $2\pi r = \pi d$ (c = *circumference*—that is, the perimeter of circle). $\overline{DE}$ is a *chord*.

NOTE

The diameter is the largest chord. ℓ_1 is a *tangent*—a line hitting a circle at only one point. ℓ_2 is a *secant*—a line hitting a circle at two points.

Sector

A *sector* is a part of a circle, like a section of a pie. θ (theta, a Greek letter like π) usually indicates an angle. s (small s) is the *arc length*, part of the circumference.

Angles are measured here in two different ways: *degrees* (as you know) and *radians*, which we will talk more about in the trig chapter. Area A = $(\theta°/360°)\pi r^2$ or $(1/2)\theta r^2$, where θ is in radians. s = $(\theta°/360°)2\pi r$ or θr, where θ is in radians. Perimeter of the sector is s + 2r. Once around a circle is 360°, or 2π radians.

OTHER SHAPES

DEFINITIONS

Trapezoid: A = $(h/2)(b_1 + b_2)$. p = $b_1 + n + b_2 + m$. b_1 is parallel to b_2. m is not parallel to n.

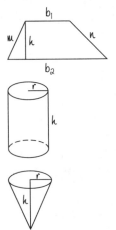

Cylinder: V = $\pi r^2 h$.

SA = πr^2 + πr^2 + $2\pi rh$ = $2\pi r(r + h)$.
         top   bottom   curved
                          surface

Cone: V = $(1/3)\pi r^2 h$. When a three-dimensional figure, such as a cone or pyramid, comes to a point, multiply the volume of the cylinder by 1/3.

Triangle: We will say a lot about them. A = (1/2)bh.
p = a + b + c.

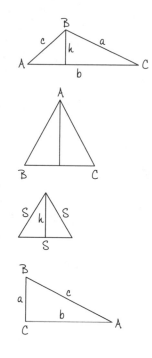

Isosceles triangle: At least two sides are equal. BC is
the base (unequal). AB and AC are the legs (equal).
Angle A is the *vertex angle* (unequal). Angles B and C
are the *base angles* (equal).

Equilateral triangle: All sides are equal. All angles are
60° since the sum of the angles of a triangle is 180°.
p = 3s. h = $s3^{1/2}/2$. A = $s^2 3^{1/2}/4$.

Right triangle: a and b are the *legs* (may be equal). c is
the *hypotenuse* (opposite right angle). You should
know certain right triangles perfectly. Also, $c^2 = a^2 + b^2$.
For 30-60-90° triangles, sides s, $s3^{1/2}$, 2s. Specifically, if
the hypotenuse is 2, opposite the 30° angle is 1, and
opposite the 60° side is $3^{1/2}$. For 45-45-90° triangles,
sides s, s, $s2^{1/2}$. Specifically, if opposite 45° is 1, the
hypotenuse is $2^{1/2}$.

All of these are needed for trig, as well as the
following groups: 3, 4, 5 (the largest is always the
hypotenuse); 6, 8, 10; 9, 12, 15, etc. (for this one, since
it appears in physics, you might learn the angles—37°
approximately, 53° approximately, 90° exactly);
5, 12, 13 (10, 24, 26); 8, 15, 17; 7, 24, 25. There are
others, but it pays to learn these because they come up
over and over and over and over.

OTHER DEFINITIONS

Supplementary: Two angles whose sum is 180°.

Complementary: Two angles whose sum is 90°. (Note
the spelling—*complimentary* means "How attractive all
you readers are," a compliment. Also note: I hate measures of angles, for those of you who know what that is.)

NOTE

This chapter is on
the 21st century
MATH SAT.

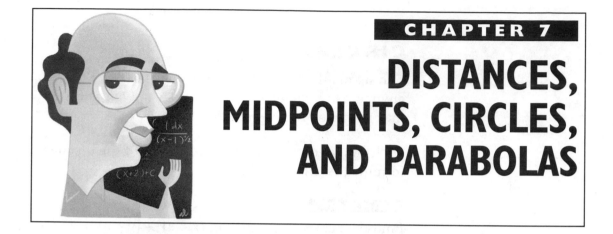

CHAPTER 7

DISTANCES, MIDPOINTS, CIRCLES, AND PARABOLAS

DISTANCES

The *distance* between two points (x_1, y_1) and (x_2, y_2) is given by the formula $d = [(x_2 - x_1)^2 + (y_2 - y_1)^2]^{1/2}$.

MIDPOINTS

The *midpoint*—that is, the average (mean) between two points—is given by the formula $((x_1 + x_2)/2, (y_1 + y_2)/2)$.

NOTE

The chapter is on the 21st century MATH SAT.

EXAMPLE 1

Given $(3, -5)$ and $(9, -1)$. Find the length of the line segment between the points and the midpoint of that line segment.

$d = [(9 - 3)^2 + (-1 - (-5))^2]^{1/2} = (36 + 16)^{1/2} = 52^{1/2}$

$\quad = (2 \cdot 2 \cdot 13)^{1/2} = 2(13)^{1/2}$

The midpoint is $[(3 + 9)/2, (-1 + -5)/2] = (6, -3)$.

CIRCLES

The set of all points (x, y) at a distance r (don't tell anyone, but r stands for *radius*) from a given point (h, k) is called the *center of the circle*.

The equation of a circle is the square of the distance formula $(x - h)^2 + (y - k)^2 = r^2$.

EXAMPLE 2—

Find the center and radius of $(x - 3)^2 + (y + 5)^2 = 7$.

$C = (3, -5)$, which is the opposite sign, and the radius $= 7^{1/2}$.

EXAMPLE 3—

Find r and C for the circle $2x^2 + 2y^2 - 12x + 6y + 8 = 0$.

First, we know it's a circle since the coefficients of x^2 and y^2 are the same, as long as the radius turns out to be a positive number.

Coefficients of x^2 and y^2 must be 1.	$2x^2 + 2y^2 - 12x + 6y + 8 = 0$ or $x^2 + y^2 - 6x + 3y + 4 = 0$
Group x's and y's together, constant to other side.	or $x^2 - 6x + y^2 + 3y = -4$
Complete the square and add the term(s) to each side.	or $x^2 - 6x + (-6/2)^2 + y^2 + 3y + (3/2)^2 = (-6/2)^2 + (3/2)^2 - 4$
Factor and do the arithmetic.	or $(x - 3)^2 + (y + 3/2)^2 = 29/4$

The center is $(3, -3/2)$, and the radius is $29^{1/2}/2$.

PARABOLAS

Later we will do a more complete study of the parabola. We need the basic, standard high school parabola for now. We will study parabolas of the form

$y = ax^2 + bx + c$, $a \neq 0$. The coefficient of x^2 determines the parabola's shape.

The low point (or high point), indicated by the letter V, is the *vertex*, which is why the letter V is usually used.

The x coordinate of the vertex is found by setting x equal to $-b/(2a)$. The y value is gotten by putting the x value into the equation for the parabola. The line through the vertex, the *axis of symmetry*, is given by $x = -b/(2a)$. The intercepts plus the vertex usually are enough for a fairly good picture. We will do some now.

EXAMPLE 4—

$y = 2x^2 - 7x + 3$

Vertex $x = \dfrac{-b}{(2a)} = \dfrac{-(-7)}{2(2)} = \dfrac{7}{4}$

$y = 2\left(\dfrac{7}{4}\right)^2 - 7\left(\dfrac{7}{4}\right) + 3 = \dfrac{-25}{8}$ $\left(\dfrac{7}{4}, \dfrac{-25}{8}\right)$

Axis of symmetry: $x = \dfrac{-b}{(2a)}$ $x = \dfrac{7}{4}$

y intercept: $x = 0$, $y = 3$ $(0, 3)$

x intercepts: $y = 0$

$2x^2 - 7x + 3 = (2x - 1)(x - 3) = 0$

$x = 1/2, 3$, and the intercepts are $(1/2, 0)$ and $(3, 0)$. The graph opens upward.

EXAMPLE 5—

$y = x^2 + 6x$

Vertex $x = \dfrac{-b}{(2a)} = \dfrac{-6}{2} (1) = -3$

$y = (-3)^2 + 6(-3) = -9$ $(-3, -9)$

x intercepts: $y = 0$

$x^2 + 6x = x(x + 6) = 0$

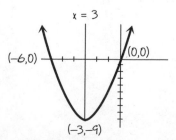

x = 0, –6—(0, 0); also, the y intercept is (–6, 0). Picture is up.

Axis of symmetry: x = –3

EXAMPLE 6—

$y = 9 - x^2$

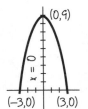

Vertex $x = \dfrac{-b}{(2a)} = \dfrac{0}{2} (-1) = 0$

$y = 9$ (0, –9) also y intercept

x intercepts: y = 0

$0 = 9 - x^2 = (3 - x)(3 + x)$

x = 3, –3; intercepts are (3, 0) and (–3, 0). Picture is down.

Axis of symmetry: x = 0

EXAMPLE 7—

$y = x^2 - 2x + 5$

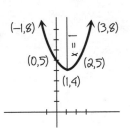

Vertex x = –(–2)/2(1) = 1

$y = 1^2 - 2(1) + 5$ (1, 4)

The y intercept is (0, 5). $y = x^2 - 2x + 5 = 0$. The quadratic formula gives imaginary roots, so no x intercepts.

To get more points, make a chart. Take two or three x values (integers) just below the vertex and two or three just above:

x	y
–1	8
0	5
1	4
2	5
3	8

Axis of symmetry: x = 1. Parabola is up.

We are now ready for functions.

CHAPTER 8

FUNCTIONS

One of the most neglected topics in high school is the study of functions. In this book there are three rather lengthy chapters directly related to functions and several others that are indirectly related. There are two reasons for this: Functions are important, and most calculus courses assume you know this topic almost perfectly, an unrealistic assumption. So let's get started at the beginning.

DEFINITION

Function: Given a set D. To each element in D, we assign one and only one element.

EXAMPLE I—

Does the picture here represent a function?

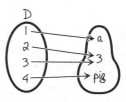

The answer is yes. 1 goes into a, 2 goes into 3, 3 goes into 3, and 4 goes into *pig*. Each element in D is assigned one and only one element.

   The next example will show what is not a function. But let us talk a little more about this example. The set D is called the *domain.* We usually think about x values when we think about the domain. This is not necessarily true, but it is true in nearly all high school and college courses, so we will assume it.

There is a second set that arises. It is not part of the definition. However, it is always there. It is called the *range.* Notice that the domain and the range can contain the same thing (the number 3) or vastly different things (3 and *pig*). However, in math, we deal mostly with numbers and letters. The rule (the arrows) is called the *map* or *mapping.* 1 is mapped into a; 2 is mapped into 3; 3 is mapped into 3; and 4 is mapped into *pig.*

FUNCTIONAL NOTATION

The rule is usually given in a different form: f(1) = a (read "f of 1 equals a"); f(2) = 3; f(3) = 3; and f(4) = *pig.*

NOTE 1
When we think of the range, we will think of the y values, although again this is not necessarily true.

NOTE 2
We cannot always draw pictures of functions, and we will give more realistic examples after we give an example of something that is not a function.

EXAMPLE 2—
The picture here does not represent a function since 1 is assigned two values, a and d.

EXAMPLE 3—
Let f(x) = x^2 + 4x + 7. D = {1, −3, 10}.

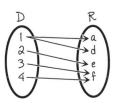

NOTE

This notation is on the 21st century MATH SAT.

$f(1) = (1)^2 + 4(1) + 7 = 12$

$f(-3) = (-3)^2 + 4(-3) + 7 = 4$

$f(10) = (10)^2 + 4(10) + 7 = 147$

The range would be {4, 12, 147}. If we graphed these points, we would graph (1, 12), (-3, 4), and (10, 147).

NOTE

Instead of graphing points (x, y), we are graphing points (x, f(x)). For our purposes, the notation is different, but the meanings are the same.

EXAMPLE 4—

Let $g(x) = x^2 - 5x - 9$. D = {4, 0, -3, a^4, x + h}. Find the elements in the range.

This is a pretty crazy example, but there are reasons to do it.

$g(4) = (4)^2 - 5(4) - 9 = -13$

$g(0) = 0^2 - 5(0) - 9 = -9$

$g(-3) = (-3)^2 - 5(-3) - 9 = 15$

$g(a^4) = (a^4)^2 - 5a^4 - 9 = a^8 - 5a^4 - 9$

$g(x + h) = (x + h)^2 - 5(x + h) - 9$
$= x^2 + 2xh + h^2 - 5x - 5h - 9$

Wherever there is an x, you replace it by x + h!

The range is {-13, -9, 15, $a^8 - 5a^4 - 9$, $x^2 + 2xh + h^2 - 5x - 5h - 9$}.

EXAMPLE 5—

$f(x) = x/(x + 5)$. Find

$$\frac{f(x + h) - f(x)}{h}$$

Add the fractions. Two tricks: a/b − c/d = (ad − bc)/bd; (e/f)/h = e/fh.

$$\frac{f(x+h)-f(x)}{h} = \frac{\dfrac{x+h}{x+h+5} - \dfrac{x}{x+5}}{h}$$

Multiply out the top; never multiply out the bottom.

$$= \frac{(x+h)(x+5) - x(x+h+5)}{(x+h+5)(x+5)h}$$

Cancel the h's.

$$= \frac{5h}{(x+h+5)(x+5)h} = \frac{5}{(x+h+5)(x+5)}$$

This kind of problem occurs in almost every precalc book. What you should ask is why the heck it is here. I will tell you. This is very close to the first topic you deal with in calculus. Here is a preview.

We have learned that the slope of a straight line is always the same. However, if we draw any curve and draw all its tangent lines, the slope changes. We would like to study this and algebraize it.

Given the point P(x, f(x)). A little bit away from P is point Q. Its x value is x + h, where x + h is an x value a little bit away from x. (In order for you to see it, Q is far away from P.) If the first coordinate is x + h, the second coordinate is f(x + h). Draw $\overline{PQ}$, $\overline{PR}$ (horizontal line), and $\overline{QR}$ (vertical line). On any horizontal line all y values are the same. P and R have the same y values. Q and R have the same x values.

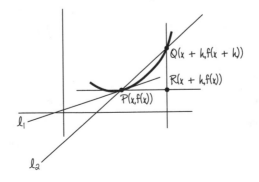

Since Q and R have the same x values, the length of $\overline{QR}$, which is the change in y or Δy, is $f(x + h) - f(x)$. Since P and R have the same y values, the length of $\overline{PR}$, which is the change in x or Δx, is $(x + h) - x = h$.

The slope of the secant line ℓ_2 joining the points P and Q is

$$\frac{\Delta y}{\Delta x} = \frac{f(x + h) - f(x)}{h}$$

which is why we study this expression. But here's the conclusion. If we let h go to 0, graphically it means the point $(x + h, f(x + h))$ gets closer and closer to $(x, f(x))$. If we do this process to the left of P as well as here to the right of P, and if they both approach the line ℓ_1, then what we have calculated is the slope of the tangent line ℓ_1 at the point $(x, f(x))$!!!!! You have taken your first step into calculus!!!!!

DEFINITION

1 – 1 function: 1 – 1 is a property we need occasionally.

If $f(a) = f(b)$, then $a = b$.

EXAMPLES 6 AND 7—

$f(x) = 2x$ is 1 – 1, but $g(x) = x^2$ is not 1 – 1.

If $f(a) = f(b)$, then $2a = 2b$ and $a = b$.

If $g(a) = g(b)$, then $a^2 = b^2$. But a could equal b or −b and therefore not 1 – 1.

How to tell a function by sight: Use the vertical line test. If, for every vertical line, each line hits the curve once and only once, then we have a function (for each x value, there is only one y value). If there is one vertical line that hits a curve twice, the curve is not a function.

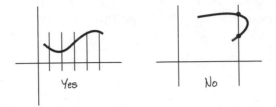

Yes   No

Finding the Ranges and Domains and Sketches of Functions (Beginning)

EXAMPLE 8—

f(x) = 3x + 1

This should look familiar. It looks like y = 3x + 1. If you graph it, you will find that it is. It is an *oblique* (slanted) *straight line*. Any such line has a domain and range, all real numbers. Let's graph it:

x	f(x)	
−1	−2	(−1, −2)
0	1	(0, 1)
2	7	(2, 7)

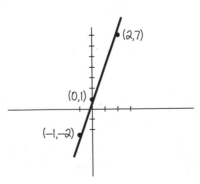

EXAMPLE 9—

$f(x) = x^2 + 6x + 5$

This is, of course, the parabolic function we did before. Let's do it once more:

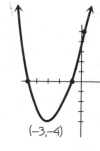

Vertex: $x = -b/(2a) = -6/2(1) = -3$. $f(-3) = (-3)^2 + 6(-3) + 5 = -4$. $(-3, -4)$.

y intercept: $x = 0$. $f(0) = (0)^2 + 6(0) + 5 = 5$. $(0, 5)$.

$(-3,-4)$

x intercepts: $f(x) = 0 = x^2 + 6x + 5 = (x + 5)(x + 1)$. $x = -5, -1$. $(-5, 0), (-1, 0)$.

Let's do some functions that are not repeats.

EXAMPLE 10—

$f(x) = |3x + 5|$

In graphing an absolute value, we should note that the shape is a $\vee$ or $\wedge$ if there is a minus sign in front of the absolute value. We find the vertex first by setting $3x + 5 = 0$ or $x = -5/3$. We then make a chart, taking two x values (only integers to make life easier for us) less than $-5/3$ and two greater than $-5/3$. Then we graph:

x	$\lvert 3x + 5 \rvert$
-3	4
-2	1
$-5/3$	0
-1	2
0	5

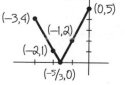

The domain is all real numbers. The range is all numbers greater than or equal to 0.

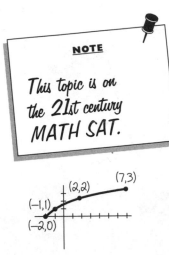

NOTE

This topic is on the 21st century MATH SAT.

EXAMPLE 11—

$g(x) = \sqrt{x + 2}$

The domain is $x + 2 \geq 0$ or $x \geq -2$. The range is $y \geq 0$ (positive square root). The object is to pick values so that the square root is exact. The values $-2, -1, 2,$ or 7 will do the trick:

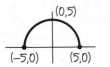

x	$\sqrt{x + 2}$
-2	0
-1	1
2	2
7	3

EXAMPLE 12—

$F(x) = \sqrt{25 - x^2}$

The domain is $25 - x^2 = (5 - x)(5 + x) \geq 0$. If we do the inequalities as before, we get the domain $-5 \leq x \leq 5$. However, we can use a trick. If we square both sides and get the x^2 on the left, we get $x^2 + y^2 = 25$, a circle center at the origin, radius 5. However, since $y \geq 0$, it is only the top half of the circle.

From the graph here, we see that the range is $0 \leq y \leq 5$.

EXAMPLE 13—

$f(x) = \dfrac{x}{|x|}$

This is an example that is needed for the future. For any value bigger than 0, no matter how close to 0, f(positive) = 1. Similarly, f(negative) = −1. f(0) is undefined. Domain is all x except 0. Range is $\{1, -1\}$.

The open dot means exactly as before. $(0, -1)$ and $(0, 1)$ are not part of the answer, but everything up to that point is.

We next get to a group of functions that almost everyone has trouble with. Please study this example carefully. We also need two new related notations:

DEFINITION (OF 2–2)

2^+: A number close (very close) to 2, bigger than 2, such as 2.00001.

2^-: A number very close to 2 but smaller than 2, such as 1.9999999.

EXAMPLE 14—

We are now ready for this example, a function defined in pieces.

A. $f(x) = x^2$ $x < 0$

B. $= 6$ $x = 0$

C. $= x + 2$ $0 < x < 2$

D. $= 7$ $x = 2$

E. $= 6 - x$ $2 < x \le 4$

F. $= 2$ $x > 4$

There are six parts. A is half a parabola. B is the point (0, 6). C is a line segment minus the two ends. D is the point (2, 7). E is another line segment minus the left end. F is a ray minus the left end.

Let's try some points. From A, $f(-3) = 9$, $f(-2) = 4$, $f(-1) = 1$, and $f(0^-) = 0^+$—an open dot on the point (0, 0). From B, $f(0) = 6$. That's it. From C, $f(0^+) = 0^+ + 2 = 2^+$, an open dot on (0, 2). $f(1) = 3$. $f(2^-) = 4^-$. From D, $f(2) = 7$. From E, $f(2^+) = 6 - 2^+ = 4^-$ (6 minus a little more than 2 is a little less than 4)—an open dot on (2, 4). $f(3) = 3$. $f(4) = 2$ (no need for 4^-). Finally, from (F), $f(4^+) = 2$ (note that E and F come together), $f(5) = 2$, $f(6) = 2$, forever $= 2$.

Let us put this in chart form and graph this function:

x	f(x)
−3	9
−2	4
−1	1
0⁻	0⁺
0	6
0⁺	2⁺
1	3
2⁻	4⁻
2	7
2⁺	4⁻
3	3
4	2
4⁺	2
5	2
6	2

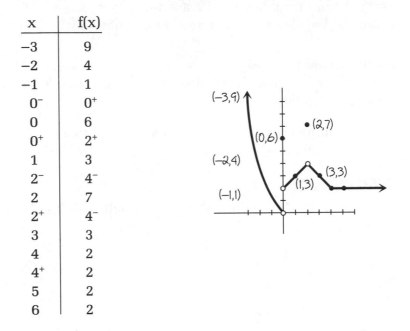

This is not easy, but it is important. This example should be gone over a good number of times.

After the first edition went to print, I became acquainted with a second way to do this problem. By doing both ways in class, my students learned how to do these kinds of important problems much more easily.

EXAMPLE 14 REVISITED

$f(x) = x^2$, but we are only interested in the part where $x < 0$. Its picture looks like this:

$f(0) = 6$. It is an isolated point. So is $f(2) = 7$. Their pictures look like this:

F(x) = x + 2, but we are interested only in the graph
between 0 and 2, not including either end. Its picture
looks like this:

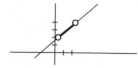

f(x) = 6 − x between 2 and 4, including 4, excluding 2.
Its picture looks like this:

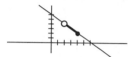

f(x) = 2, but only if x > 4. Its picture looks like this:

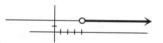

Put them all together and it looks like this:

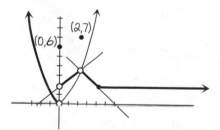

Read both of these a couple of times and you should
be OK.

COMPOSITE FUNCTIONS

Suppose we have a function map f whose domain is D
and whose range is R_1. Suppose also we have another

function map g, domain R_1 and range R_2. Is there a map that goes from the original domain D to the last range R_2? The answer is, of course, yes; otherwise why would I waste your time writing this paragraph?

DEFINITION

Given a function map f, domain D, range R_1. Given a second function map g, domain R_1, range R_2. Define the composite map g ∘ f, domain D, range R_2. (g ∘ f)(x)— read "g circle f of x"—is g[f(x)], read "g of f of x." The picture might look as shown here:

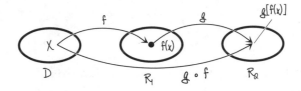

EXAMPLE 15—

Suppose $f(x) = x^2 + 4$ and $g(x) = 2x + 5$.

A. g[f(3)]. $f(3) = 3^2 + 4 = 13$. $g(13) = 2(13) + 5 = 31$. The picture is as shown here:

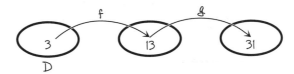

B. f[g(3)]. First, let's note that the picture would be totally different. Second, the "inside" map always is done first. $g(3) = 2(3) + 5 = 11$. $f(11) = 11^2 + 4 = 125$. Rarely are the two composites the same.

C. g[f(x)]. $g(x) = 2x + 5$. This means multiply the point by 2 and add 5. g[f(x)] means the point is no longer x but f(x). The rule is multiply f(x) by 2 and add 5. So g[f(x)] = 2[f(x)] + 5 = 2(x² + 4) + 5 = 2x² + 13.

NOTE

g[f(3)] = 2(3)² + 13 = 31, which agrees with item A above.

D. f[g(x)]. f(x) = x² + 4. So f[g(x)] = [g(x)]² + 4 = (2x + 5)² + 4 = 4x² + 20x + 29. f[g(3)] = 4(3)² + 20(3) + 29 = 125, which agrees with item B, as it must.

INVERSE FUNCTIONS

There is a function that takes the domain into the range. Is there a function that takes the range back into the domain? The answer is yes under certain circumstances. That circumstance is when the domain and the range are in 1:1 correspondence. This means every element in the domain can be paired off with one and only one element in the range. In the case of a finite set, 1:1 correspondence means the same number of elements.

On a graph you can see whether you have an inverse by first using the vertical line test (each line hits the graph only once) to see whether the curve is a function. Then use the horizontal line test (each line also hits the curve only once) to determine whether there is an inverse function. The most common kinds of functions that have inverses are ones that always increase or always decrease.

The definition of an *inverse function* is very long, but it is not very difficult after you read it and the examples that follow.

Given a function map f, domain D, range R, D and R are in 1:1 correspondence. Define f⁻¹, read "f inverse," domain R, range D. If originally a was an element of D and b was an element of R, define f⁻¹(b) = a if originally f(a) = b.

This looks quite bad, but really it is not. Look at the picture here:

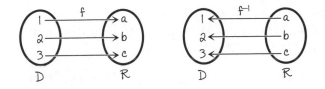

$f^{-1}(a) = 1$ because originally $f(1) = a$; $f^{-1}(b) = 2$ since $f(2) = b$; and $f^{-1}(c) = 3$ because $f(3) = c$.

A little better? Let's make it much better. First note the inverses we know: adding and subtracting, multiplying and dividing (except with 0), squaring and square roots, providing only one square root, cubing and cube roots, etc.

EXAMPLE 16—

Given map $f(x) = 2x + 3$, domain $\{1, 8, 30\}$.

$f(1) = 5$, $f(8) = 19$, $f(30) = 63$. The range is $\{5, 19, 63\}$. D and R are in 1:1 correspondence. Let's find the inverse. The domain and range switch. The new domain is $\{5, 19, 63\}$. Let's find the new map, f^{-1}. $f(x) = 2x + 3$ means multiply by 2 and add 3. Going in the opposite direction, adding 3 was last so subtracting 3 is first. First we multiply by 2; going backward, the last thing is dividing by 2.

$f(x) = 2x + 3$. $2x + 3 = f(x)$. $2x = f(x) - 3$. $x = [f(x) - 3]/2$. Notation change: $f(x)$ is now in the new domain. Change $f(x)$ to x! x is the new function; so x becomes $f^{-1}(x)$. So our new function is $f^{-1}(x) = (x - 3)/2$, D = $\{5, 19, 63\}$. Let's check it out!! $f^{-1}(5) = (5 - 3)/2 = 1$. OK so far. $f^{-1}(19) = (19 - 3)/2 = 8$. $f^{-1}(63) = (63 - 3)/2 = 30$. Everything is OK.

Let's do some more examples.

EXAMPLE 17—

$$f(x) = \frac{x + 2}{x - 3}$$

The domain is all real numbers except $x = 3$ since the bottom of a fraction may never be equal to 0. However, it is not clear at all what the range is. Let's find the inverse function. Perhaps the original range and the new domain will become more obvious. In order to make this less messy, let $f(x) = y$.

Multiply through by $x - 3$.

$$\frac{y}{1} = \frac{x + 2}{x - 3}$$

$$y(x - 3) = x + 2$$

$$yx - 3y = x + 2$$

$$yx - x = 3y + 2$$

$$x(y - 1) = 3y + 2$$

$$x = \frac{3y + 2}{y - 1}$$

At this point it is clear that the old range was all y values except for 1. Now the notation changes.

$$f^{-1}(x) = \frac{3x + 2}{x - 1}$$

Note: Sometimes, in order to find the range, it is a good idea to solve for x!!!!

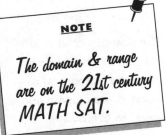

EXAMPLE 18—

Find the inverse map for

$$f(x) = \sqrt{3x + 1}$$

with the domain $x \geq 1$ and the range $y \geq 2$.

Let $f(x) = y$.

$$y = \sqrt{3x + 1}$$

$$\sqrt{3x + 1} = y$$

$$3x + 1 = y^2$$

$$3x = y^2 - 1$$

$$x = \frac{(y^2 - 1)}{3}$$

so $f^{-1}(x) = \dfrac{(x^2 - 1)}{3}$, $x \geq 2$

As we go along in the book, we will have more inverses. That's it for now.

CHAPTER 9

LINEAR SYSTEMS AND MORE ABOUT $y = x^2$

EQUATIONS IN TWOS AND THREES

I have been very disappointed in the ways two equations in two unknowns and three equations in three unknowns have been presented in other books. So I've included this section.

TWO EQUATIONS IN TWO UNKNOWNS

There are many ways to solve two equations in two unknowns. Almost all of them are easy. We will show three ways and then relate the algebra to the graphing.

Graphing

EXAMPLE 1—

Solve for x and y using graphing:

$2x + y = 8$

$x - 3y = -3$

NOTE

The beginning of this chapter has always been on the *SAT MATH* but the problems are mostly done differently. (See *SAT MATH For the Clueless*.)

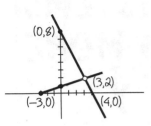

The easiest way to graph a line is using the intercept-intercept method.

$2x + y = 8$

y intercept means $x = 0$ $y = 8$ $(0, 8)$

x intercept means $y = 0$ $2x = 8$ $x = 4$ $(4, 0)$

$x - 3y = -3$ $x = 0$ $-3y = -3$ $y = 1$ $(0, 1)$

$y = 0$ $x = -3$ $(-3, 0)$

Graphing, we see that the lines meet at the point $(3, 2)$. Checking in both equations, $2(3) + 2 = 8$ annnd $3 - 3(2) = -3$.

NOTE 1

This is the most inaccurate way of solving since the intercepts and the solutions don't have to be integers, but this way might be the only way to do the problem.

NOTE 2

We will get back to the relationship between graphing and algebra later.

Algebra: Substitution

EXAMPLE 2—

Solve for x and y by substitution:

$4x + 3y = 13$

$x + 5y = 16$

We solve for x in the second equation and substitute in the first. We can do this because when the graphs meet, their x values and y values are equal:

$x = 16 - 5y$

$4(16 - 5y) + 3y = 13$

$64 - 20y + 3y = 13$

$-17y = -51$

$y = 3$

$x = 16 - 5y = 16 - 5(3) = 1$

The point is (1, 3).

This looks pretty easy, but sometimes it is messier.

EXAMPLE 3—

Solve for x and y by substitution:

$2x + 5y = 5$

$3x - 2y = 17$

Which letter should we solve for? There is no good letter since each variable has a coefficient not equal to 1 or −1 and will result in fractions. Let's solve for x in the second equation:

$x = \dfrac{(2y + 17)}{3}$

Substitute it in the first equation.

$2\left[\dfrac{(2y + 17)}{3}\right] + 5y = 5$

Multiply by 3 to get rid of the fraction.

$2(2y + 17) + 15y = 15$

$4y + 34 + 15y = 15$

$19y = -19 \qquad y = -1$

Sooo $x = \dfrac{[2(-1) + 17]}{3} = 5$. The answer is (5, −1).

Although this is not the easiest method, it probably is used the most. However, the easiest method, for me anyhow, is . . .

Algebra: Elimination

EXAMPLE 4—

Solve for x and y by elimination:

$$4x + y = 10$$

$$x - y = 5$$

We just add, eliminating y!

$$5x = 15 \qquad x = 3$$

Substituting, we get $y = -2$. $(3, -2)$.

Well, that was easy. Let's try another.

EXAMPLE 5—

Solve for x and y by elimination:

$$2x + 3y = 12$$

$$2x + y = 8$$

We just subtract, eliminating x:

$$2y = 4 \qquad y = 2$$

Substituting, $2x + 2 = 8$. $2x = 6$. $x = 3$. The solution is $(3, 2)$.

Are they all this easy? Yes, pretty easy.

EXAMPLE 6—

Solve for x and y by elimination:

$$2x + 5y = 5$$

$$3x - 2y = 17$$

Multiply the top equation by 3 and the bottom by −2 and add:

$3(2x + 5y = 5)$ $6x + 15y = 15$

$−2(3x − 2y = 17)$ $−6x + 4y = −34$

$$19y = −19$$

$$y = −1$$

Now let's eliminate y by multiplying the top equation by 2 and the bottom by 5 and adding:

$2(2x + 5y = 5)$ $4x + 10y = 10$

$5(3x − 2y = 17)$ $15x − 10y = 85$

$$19x = 95$$

$$x = 5$$

The solution is (5, −1).

NOTE 1

We could have substituted $y = −1$ into either equation. However, if $y = 37/83$, we would not like to substitute. One of the great things about this method is that no matter how messy the numbers, the problem is done easily by double elimination.

NOTE 2

To eliminate either letter, multiply by the number that gives the LCD (lowest common denominator). 2x and 3x: LCD is 6. 2x by 3 and 3x by −2. If you don't want to think, just use the opposite number; in this problem, both methods are the same.

NOTE 3

2x and 3x: both had the same sign, so we multiplied one equation by a + sign and one by a − sign to add. It didn't matter which by which.

NOTE 4

+5y and −2y had different signs. We multiplied both by a plus. If we were crazy, we could have used two minus signs.

NOTE 5

If we have fractions, clear fractions by multiplying each equation by its LCD (it doesn't have to be the same) and proceed.

NOTE 6

If the letters and/or numbers are not lined up, line them up and proceed.

Relating the Graph to the Algebra

EXAMPLE 7—

$x + y = 6$

$x − y = 2$

Adding, $2x = 8$ $x = 4$. Substituting, $y = 2$. The graph is

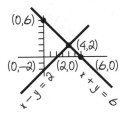

If the lines meet, we get a unique point, one value for x and one for y.

EXAMPLE 8—

Multiply by −3. $x + y = 2$

Leave alone. $−3x − 3y = −6$

                                        $3x + 3y = 12$

Add. $0 = 6$

When does $0 = 6$? *Neverrrrr!* Let us see what this means graphically:

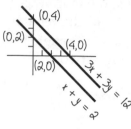

We see that the lines are parallel. They never meet. There is no solution!!!

EXAMPLE 9—

$x + 2y = 4$	$-2x - 4y = -8$	**Multiply by –2.**
$2x + 4y = 8$	$2x + 4y = 8$	**Leave alone.**

Adding, we get $0 = 0$. When does $0 = 0$? Always!! Let's see what this means graphically:

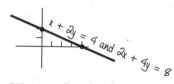

We see that both lines are the same line. The whole line is the answer, an infinite set.

In solving two equations in two unknowns, there are three possibilities: no answer (parallel lines); one point answer (the lines meet); and an infinite number of answers (both lines are the same line). What about three equations in three unknowns?

THREE EQUATIONS IN THREE UNKNOWNS

We will solve three equations in three unknowns by using the elimination method only. Let's show why graphing won't be used.

Graphing a Point in 3D

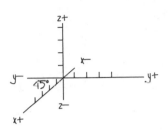

Since we now have three unknowns, the graphing must be in 3D. We draw the y and z axes the regular way. The x axis is out of the board, at a 45° angle at 70% $(1/2^{1/2})$ of the markings of the other two axes.

EXAMPLE 10—

Graph the point (3, 4, 5).

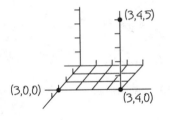

We go three units in the x direction—the point (3, 0, 0). Now four units in the y direction, at an angle, of course, to the point (3, 4, 0). Now five units up to the point (3, 4, 5). You can see why we will never graph a 3D graph point by point.

EXAMPLE 11—

Graph the plane $2x + 3y + 4z = 12$.

We will graph using the intercepts:

x = 0   y = 0   z = 3   (0, 0, 3)

x = 0   z = 0   y = 4   (0, 4, 0)

y = 0   z = 0   x = 6   (6, 0, 0)

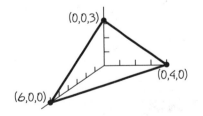

This is how you graph one plane on a graph. Can you imagine graphing three on the same set of axes and telling where they meet? Maybe you are superperson, but not me.

What Would It Look Like?

If we could graph three planes on the same graph, what would it look like?

1. It could be a point. Look at the corner of a room.

2. It could be a line. Look at the binding of a book.

3. It could be a plane. The three equations are the same plane.

4. There could be no solutions. A number of possibilities exist: three parallel planes, two parallel planes and one plane intersecting both parallel planes, three planes each intersecting the other two but not at the same place (like a 3D triangle), or two planes the same and the other parallel.

We will do only the problem where you get one point. (Planes are not necessarily perpendicular.)

EXAMPLE 12—

Solve for x, y, and z:

A. $4x + 2y + z = 12$

B. $2x - 3y - z = -8$

C. $3x + 2y + 2z = 15$

We will eliminate the easiest letter. In this case, it is z.

Add A and B.

Add twice B to C.

A + B.
$$4x + 2y + z = 12$$
$$2x - 3y - z = -8$$
$$\text{D.} \quad 6x - y = 4$$

2B + C.
$$2(2x - 3y - z) = 12(2)$$
$$1(3x + 2y + 2z) = 15$$
$$\text{E.} \quad 7x - 4y = -1$$

Now we solve two equations in two unknowns:

D. $6x - y = 4$

E. $7x - 4y = -1$

Multiply D by -4 and leave E alone to eliminate y, the easier letter:

$-4(6x - y) = 4(-4)$

$1(7x - 4y) = -1$

F. $-17 \ x = -17 \quad x = 1$

Substitute $x = 1$ into D or E:

$6(1) - y = 4 \qquad -y = -2 \qquad y = 2$

Substitute $x = 1$ and $y = 2$ into A, B, or C:

$4(1) + 2(2) + z = 12 \qquad 8 + z = 12 \qquad z = 4$

The answer is (1, 2, 4). Whew.

NOTE

The first thing you must do is make sure you copied the problem correctly. No kidding! If one thing is wrong, the numbers could be sooo terrible you might throw up. Be very, very careful!

ANOTHER NOTE

z was the easiest number to eliminate since you could add two equations and eliminate z. y was almost as easy. x was the worst here.

LAST NOTE

Some of the notes on two equations in two unknowns may be needed.

Later we will do two nonlinear equations in two unknowns. At least one and usually both will not

be lines. Some of the techniques are the same,
buutt . . .

y = x², y = x³,
AND A LITTLE MORE

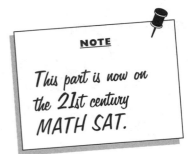

Finally, there are two functions we should know verrry
well since they come up over and over and over again.

EXAMPLE 13—

$y = x^2$

We have seen this before. Vertex is at (0, 0). Its picture
looks like this:

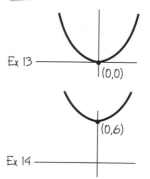

Ex 13 ——— (0,0)

EXAMPLE 14—

$y = x^2 + 6$

All the y values add 6. Vertex is (0, 6).

(0,6)

Ex 14 ———

EXAMPLE 15—

$y = x^2 - 4$

All y values subtract 4. Vertex is (0, –4).

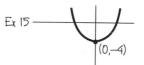

Ex 15 ———

(0,–4)

EXAMPLE 16—

$y = (x - 5)^2$

The picture moves five units to the right. Vertex is (5, 0).

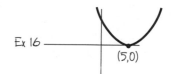

Ex 16 ———

(5,0)

EXAMPLE 17—

$y = (x + 2)^2$

This one moves two units to the left. Vertex is (–2, 0).

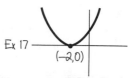

Ex 17 ———

(–2,0)

NOTE
Examples 14 through 17 are called *translations*. We
will see them again now and several times later. To
continue . . .

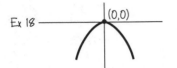

Ex 18

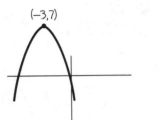

EXAMPLE 18—

$y = -x^2$

Curve is upside down. Vertex is again $(0, 0)$.

EXAMPLE 19—

$y = -(x + 3)^2 + 7$

This is a combo. Upside down, shift three units to the left, shift seven units up. Vertex is $(-3, 7)$.

EXAMPLE 20—

Draw $y = \frac{1}{4}x^2$, $y = x^2$, and $y = 4x^2$ on the same graph.

The smaller the coefficient (smaller positive), the wider the parabola. The larger the coefficient, the faster the parabola rises:

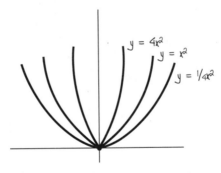

Let's look at x^3 now.

EXAMPLE 21—

A. $y = x^3$

B. $y = x^3 + 4$ (shift four units up)

C. $y = (x - 3)^3$ (shift three units to the right)

D. $y = -x^3$ upside down and so on

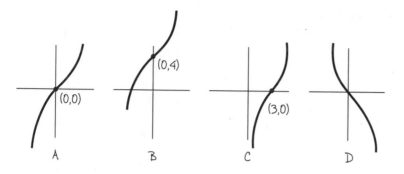

Finally, we could use this to sketch curves similar to what we know.

EXAMPLE 22—

A. $y = x^{1/2}$

B. $y = -(x - 4)^{1/2}$

A is the square root of x. We did this earlier. B is a shift of four units to the right and upside down. The pictures look like this:

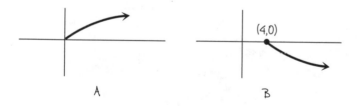

NOTE

This is not the way I prefer to teach curves such as those in Example 22, but the technique is useful, and some teachers like it. That's why I added this section.

CHAPTER 10

TRIGONOMETRY

I have always thought that trigonometry should be the easiest of all math courses. It requires very little of the past: some elementary algebra like factoring, multiplying terms, adding fractions, and solving simple equations; a little intermediate algebra; and a small amount of geometry. It does require a fair amount of memorization. I will try to convince you how easy the course should be.

ANGLES

There are three basic measures of angles: revolutions, degrees, and radians. The one we know the best is degrees. A *degree* is 1/360th of a circle. It is a purely artificial measurement, probably found only on the planet Earth in the entire universe, where hopefully there is a lot more intelligent life.

A *revolution* is once around a circle. We actually use this measurement, although you may not recognize it. The speed of a car engine or a record is measured in rpm, or revolutions per minute. A 45-rpm record is designed to make 45 revolutions per minute.

To → From ↓	rev	deg	rad
rev	1	360	2π
deg	$1/360$	1	$\pi/180$
rad	$1/2\pi$	$180/\pi$	1

The most important measurement, used in calculus because it is a pure number, is *radians*. A radian, approximately 57°, is derived by putting the length of the radius on the circumference of a circle. The central angle formed by the two radii is a radian. There are 2π radians in a circle.

Here is a convenient chart that shows how to change from one measurement to the other. Multiply the left column by the factor in the chart to give the top of the second, third, or fourth column.

EXAMPLE 1—

Change 7.5 revolutions to degrees and radians.

$7.5(2\pi) = 15\pi$ radians

$7.5(360°) = 2700°$

EXAMPLE 2—

Change 30° to radians.

$$30\left(\frac{\pi}{180}\right) = \frac{\pi}{6} \text{ radians}$$

EXAMPLE 3—

Change $\pi/4$ to degrees.

$$\left(\frac{\pi}{4}\right)\left(\frac{180}{\pi}\right) = 45°$$

You must be able to do Examples 2 and 3 very quickly. It would also be extremely beneficial if you knew the following angles: 30° = $\pi/6$; 45° = $\pi/4$; 60° = $\pi/3$; 90° = $\pi/2$; 180° = π; 270° = $3\pi/2$; 360° = 2π. The more multiples of 30°, 45°, and 60° you know, the easier things will be. Remember, I am trying to make things as easy as possible, but there is much memorization required on your part.

NOTE

Many books break down degrees. There are 60 minutes to a degree and 60 seconds to a minute. Modern treatment uses tenths and hundredths of degrees. Using minutes and seconds to me is both too old-fashioned and, even worse, too picky. Therefore, I round off all angles to the nearest degree.

BASIC DEFINITIONS

On this planet angles are measured positively in a counterclockwise direction as indicated by the picture here. We locate the point (x, y) and label r, where $r = (x^2 + y^2)^{1/2}$, the distance to the origin, which is therefore always positive. Define the following:

Sine of A =
  sin A = y/r

Cosine of A =
  cos A = x/r

Tangent of A =
  tan A = y/x

Cotangent of A =
  cot A = x/y

Secant of A =
  sec A = r/x

Cosecant of A =
  csc A = r/y

NOTE 1

When you get the point (x, y), the triangle formed must always be to the x axis, either down or up to it.

NOTE 2

You must know these definitions perfectly!!!!!

NOTE 3

The quadrant will tell you the sign of the trig function:

In quadrant I, x and y are positive (as r always is)— all trig functions are positive.

In II, x is negative—only the sine and cosecant are positive.

In III, x and y are negative—only the tangent and cotangent are positive.

In IV, y is negative—the cosine and secant are positive.

NOTE 4
You must also know Note 3 perfectly.

MULTIPLES OF 30°, 45°, 60°, AND 90°

You must be able to find all multiples of the above angles *without* a calculator. First, the three basics. Note that you do *not* have to memorize these. You should *not* memorize these. If you can draw the triangles and know the definitions, that will be enough.

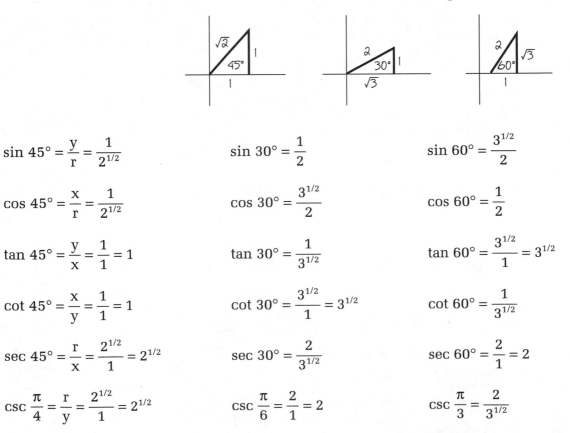

$$\sin 45° = \frac{y}{r} = \frac{1}{2^{1/2}}$$

$$\cos 45° = \frac{x}{r} = \frac{1}{2^{1/2}}$$

$$\tan 45° = \frac{y}{x} = \frac{1}{1} = 1$$

$$\cot 45° = \frac{x}{y} = \frac{1}{1} = 1$$

$$\sec 45° = \frac{r}{x} = \frac{2^{1/2}}{1} = 2^{1/2}$$

$$\csc \frac{\pi}{4} = \frac{r}{y} = \frac{2^{1/2}}{1} = 2^{1/2}$$

$$\sin 30° = \frac{1}{2}$$

$$\cos 30° = \frac{3^{1/2}}{2}$$

$$\tan 30° = \frac{1}{3^{1/2}}$$

$$\cot 30° = \frac{3^{1/2}}{1} = 3^{1/2}$$

$$\sec 30° = \frac{2}{3^{1/2}}$$

$$\csc \frac{\pi}{6} = \frac{2}{1} = 2$$

$$\sin 60° = \frac{3^{1/2}}{2}$$

$$\cos 60° = \frac{1}{2}$$

$$\tan 60° = \frac{3^{1/2}}{1} = 3^{1/2}$$

$$\cot 60° = \frac{1}{3^{1/2}}$$

$$\sec 60° = \frac{2}{1} = 2$$

$$\csc \frac{\pi}{3} = \frac{2}{3^{1/2}}$$

NOTE I

When you draw the triangles, really make them look correct (also use a ruler). In 45-45-90° triangles, make x and y equal. In the 30-60-90° triangles, make the short side (opposite the 30° angle) shorter than the longer leg (opposite the 60° angle).

NOTE 2

Your teacher might have you rationalize some of these, such as $1/2^{1/2} = 2^{1/2}/2$, by multiplying top and bottom by the square root of 2.

EXAMPLE 4—

Find cos 240°.

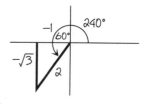

$240° = 90° + 90° + 60°$; 60° angle in the third quadrant. $3^{1/2}$ opposite the 60° angle, 1 opposite the 30° angle, and both negative (negative to the left, negative down); $r = 2$ and positive since r is always positive. $\cos 240° = \cos 4\pi/3 = x/r = -1/2$.

EXAMPLE 5—

Find sin 135°.

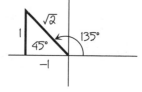

$135° = 90° + 45° = 180° - 45°$. $1 - 1 - \sqrt{2}$ triangle. x is negative, to the left; y is positive, up; $r = 2^{1/2}$ is always positive. $\sin 135° = \sin 3\pi/4 = y/r = 1/2^{1/2}$.

EXAMPLE 6—

Find tan 330°.

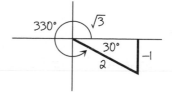

$300° = 90° + 90° + 90° + 60° = 360° - 30°$. $x = 3^{1/2}$ and is positive (to the right). $y = -1$ and is negative (down). $r = 2$ and is always positive. $\tan 330° = \tan 11\pi/6 = y/x = -1/3^{1/2}$.

NOTE

If θ is an angle in the first quadrant, the *related angles* are $180° - θ$ in the second quadrant, $180° + θ$ in the third quadrant, and 360° − θ in the fourth quadrant. The related angles all have the same picture. The value of the trig function is the same except for the sign. cos 60° = 1/2. cos 120° = cos (180° − 60°) = −1/2. cos 240° = cos (180° + 60°) = −1/2. cos 300° = cos (360° − 60°) = 1/2. These should not be memorized. You can and should draw these pictures, first to see that these are the same triangle and then to practice drawing until you become very, very good.

EXAMPLE 7—

Suppose cot A is −7/6 in quadrant IV. Write all the other trig functions of A. (This example should have preceded the 30-60-90° section.)

In IV, x is positive, y is negative, and r is positive. cot A = x/y. Let x = 7, y = −6, r = $[7^2 + (−6)^2]^{1/2}$ = $85^{1/2}$. sin A = y/r = $−6/85^{1/2}$. cos A = x/r = $7/85^{1/2}$. tan A = y/x = −6/7. cot A = x/y = −7/6. sec A = r/x = $85^{1/2}/7$. csc A = r/y = $85^{1/2}/(−6)$.

EXAMPLE 8—

sin B = 5/13 in II. Find sec B.

Let y = 5 and r = 13. $x^2 + 5^2 = 13^2$. x = $±(144)^{1/2}$ = ±12. −12 since x is negative in II. sec B = r/x = 13/−12.

NOTE

If you memorized Pythagorean triples, you would know this is 5-12-13.

ALSO NOTE

This should have also preceded the 30-60-90° section. Now we are in the correct spot.

EXAMPLE 9—

Find all the trig functions of 270°.

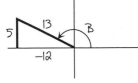

If you draw an angle of 270° or any multiple of 90°, it will fall on an axis—in this case on the negative y axis. We let r = 1 in this case. So the point will be (0, −1), x = 0 and y = −1. Now we can get all the trig functions:

(0,−1)

$$\sin 270° = \frac{y}{r} = \frac{-1}{1} = -1 \qquad \cos 270° = \frac{x}{r} = \frac{0}{1} = 0$$

$$\tan 270° = \frac{y}{x} = \frac{-1}{0}, \text{ undefined} \qquad \cot 270° = \frac{x}{y} = \frac{0}{-1} = 0$$

$$\sec \frac{3\pi}{2} = \frac{r}{x} = \frac{1}{0}, \text{ undefined} \qquad \csc 270° = \frac{r}{y} = \frac{1}{-1} = -1$$

NOTE

For all multiples of 90°, two trig functions will always be 0, two will always be undefined, and two will either both be −1 or +1.

CURVE SKETCHING

We will sketch basic sine, cosine, and tangent functions. The techniques for sketching the secant and cosecant are the same as for the sine and cosine except for a different picture, and the cotangent is likewise similar to the tangent. All of the trig functions are *periodic*. That is, after a while, they repeat. The technical definition of *periodic* is f(x + p) = f(x) for all x. If p is the smallest positive such number, p is called the *period*. The period of the sine, cosine, secant, and cosecant is 360° or 2π. The period of the tangent and cotangent is 180° or π. If we graph y = sin x by taking 0°, 30°, 45°, 60°, 90°, 120°, 135°, 150°, 180°, 210°, 225°, 240°, 270°, 300°, 315°, 330°, and 360°, the picture would look as shown here:

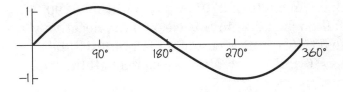

Doing the same for y = cos x, we get the second picture here:

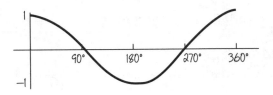

Doing the same for y = tan x, except only up to 180°, we get the third picture:

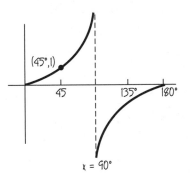

There is an *asymptote* (a line the curve gets very close to at the end but does not hit), which we will discuss later. On the tangent curve, we label one point, the 1/4 point. tan 45° = 1.

NOTE I

In sketching a curve, you will be normally asked to graph five points: the beginning, the end, and the 1/4, 1/2, and 3/4 points.

NOTE 2

A sneaky (but good) way of calculating, say, sin x at multiples of 90° is to look at the intercepts (0°, 180°, and 360°), high point, and 270°, the low point. Try it. You'll like it.

There are four things we consider in the sketching: amplitude, period, shift up and down, and phase (left and right shift). The technical definition of the *amplitude* is the maximum value minus the minimum value divided by 2. Tangent, cotangent, secant, and cosecant have no amplitude since their maximums are plus infinity and their minimums are minus infinity. For y = A sin x or y = A cos x, the amplitude is the absolute value of A.

We will work in degrees because it is easier.

EXAMPLE 10—

y = 10 sin x

The sketch is the same as y = sin x except that the high point is 10 and the low point is –10:

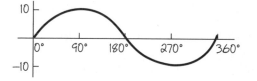

EXAMPLE 11—

y = –4 cos x

Amplitude is 4. Curve is upside down!

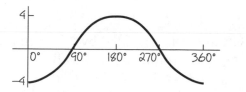

Easy so far? It really shouldn't get much harder.

EXAMPLE 12—

y = sin 5x

The amplitude is 1. The period is 360°/5 = 72°.

For sin x, the five points are 0°, 90°, 180°, 270°, and 360°. Now divide them by 5. The five points are 0°, 18°, 36°, 54°, and 72°. The curve is the sine curve:

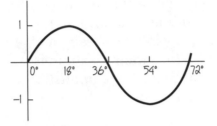

EXAMPLE 13—

y = −7 cos 2x

Amplitude 7. Upside down, period 360°/2 = 180°.

0°, 90°, 180°, 270°, and 360° divided by 2 are 0°, 45°, 90°, 135°, and 180°. Cosine curve:

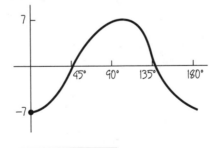

EXAMPLE 14—

y = 4 tan 3x

No amplitude. Right side up. Period 180°/3 = 60°.

Points 0°, 45°, 90°, 135°, and 180° divided by 3 become 0°, 15°, 30°, 45°, and 60°. Asymptote at the middle or 30°. The 1/4 point 15° is (15°, 4). This indicates the coefficient in front. Let us show y = 4 tan 3x = 4 tan 3(15°) = 4 tan 45° = 4(1) = 4. (15°, 4):

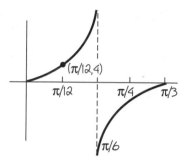

NOTE

If y = sin Bx or cos Bx, period is 360°/|B|. tan Bx?
Period is 180°/|B|.

EXAMPLE 15—

y = 5 sin (3x + 120°)

Factor the 3.

   y = 5 sin 3(x + 40°). Amplitude is 5. Right side up.
Period 360°/3 = 120°, 0°, 90°, 180°, 270°, and 360°
divided by 3 become 0°, 30°, 60°, 90°, and 120°. +40 is
the left or right shift. +40 means a shift of 40° to the left.
So we have to *subtract* 40° from each angle. 0° – 40°, 30° –
40°, 60° – 40°, 90° – 40°, and 120° – 40°, or –40°, –10°,
20°, 50°, and 80°. Let me show you why. For x = 0 – 40°,
5 sin (3x + 120°) = 5 sin [3(–40°) + 120°] = 5 sin 0 = 0:

EXAMPLE 16—

y = 12 cos (10x – 20°)

Amplitude 12. y = 12 cos 10(x – 2°).

   Period 360°/10 = 36°. The five points are 0°, 9°, 18°,
27°, and 36°. Phase shift is 2° to the *right.* 0° + 2°,

9° + 2°, 18° + 2°, 27° + 2°, 36° + 2°, or 2°, 11°, 20°, 29°, and 38°:

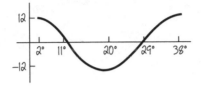

EXAMPLE 17—

$y = \sin (\tfrac{1}{2}x + 50°)$

Factor out ½. $y = \sin \tfrac{1}{2}(x + 100°)$.

    Amplitude is 1. Period is 360°/½ = 720°. Shift 100° to the left. 0°, 90°, 180°, 270°, and 360° become 0°, 180°, 360°, 540°, and 720°. Subtract 100°; we get −100°, 80°, 260°, 440°, and 620°:

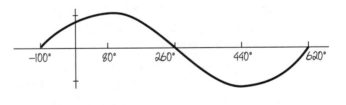

EXAMPLE 18—

$3y - 12 = 6 \sin (3x - 60°)$

$3y = 6 \sin 3(x - 20°) + 12$

$y = 2 \sin 3(x - 20°) + 4$

Note that the 3 in front of the angle is unchanged.

    Amplitude is 2. Right side up. Period is 360°/3 = 120°. Shift 20° to the right. 0°, 90°, 180°, 270°, and 360° become 0°, 30°, 60°, 90°, and 120° become 20°, 50°, 80°, 110°, and 140°. The 4 on the right is a shift of the whole graph *up 4* (add 4 to each y value).

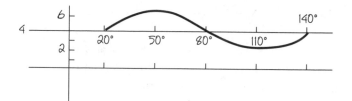

NOTE

Of course, the negative sign at the right (−4 instead of +4) would be a shift down.

If this does not look too hard, it shouldn't be. This should not take too much practice either.

IDENTITIES

This part of trig is important. It is also difficult for several reasons. Number 1, almost no book really teaches you how to attack identities. Number 2, no one method really works for all identities. Number 3, some of them require some thinking! Number 4, identities are one of the four or five areas of math that require a *lot* of practice!!!!!

First, we would like to know what an identity is.

DEFINITION

Identity: An equation that is true for every value for which it is defined.

EXAMPLE 19—

$2x + 3x = 5x$

No matter what x you substitute, $2x + 3x = 5x$.

EXAMPLE 20—

$4/x + 5/x = 9/x$

It is not defined for x = 0. Otherwise it is true.

The following are the most important trig identities (which are proven in virtually every book). In order to do this section, you must know these identities perfectly in all forms. Here is the list:

1. $\sin x \csc x = 1$. $\sin x = 1/\csc x$ and $\csc x = 1/\sin x$.

2. $\cos x \sec x = 1$. $\cos x = 1/\sec x$ and $\sec x = 1/\cos x$.

3. $\tan x \cot x = 1$. $\tan x = 1/\cot x$ and $\cot x = 1/\tan x$.

4. $\tan x = \sin x/\cos x$

5. $\cot x = \cos x/\sin x$.

6. $\sin^2 x + \cos^2 x = 1$. $\sin^2 x = 1 - \cos^2 x$ and $\cos^2 x = 1 - \sin^2 x$.

7. $1 + \tan^2 x = \sec^2 x$. $\tan^2 x = \sec^2 x - 1$.

8. $1 + \cot^2 x = \csc^2 x$. $\cot^2 x = \csc^2 x - 1$.

You *must* know these perfectly. Say I came to your house in the middle of the night and said, "What is $\tan^2 x$ equal to?" You would mumble, "It is either $\sec^2 x - 1$ or $\sin^2 x/\cos^2 x$ or $1/\cot^2 x$. Now let me go back to sleep!!!!" When you use the identities, if you have not learned them yet, keep this list in front of you. It is *very* important to learn them. Am I repeating myself??!!

Following is a list of steps you should try, in the general order you should try them. Unlike solving first-degree equations, where if you follow the steps in precisely the right (and same) order you will always get the correct answer, sometimes you must go back and sometimes you must skip steps. Sometimes, even if you do a correct step, it might not be correct for this particular problem. But many times here you learn more from a wrong step than a right one, as long as you remember for another problem. But you must keep trying. Only in that way will you get better.

Attacking Identities

1. It is usually better to work with the more complicated side.

   A. Adding trig functions is more complicated than multiplication.
   B. Later on, double and half angles are more complicated than single angles—sin 2x or cos ½x is more complicated than tan x.
   C. In most books, the left side is the more complicated.

2. Do something "obvious." (I hate that word!)

   A. Add fractions.
   B. Multiply out a term in parentheses.
   C. Square a term (a binomial).

3. If there is more than one term on the top and one term on the bottom, split the fraction. In symbols,

$$(a + b + c)/d = \frac{a}{d} + \frac{b}{d} + \frac{c}{d}$$

4. If there is a known identity, use it.

5. If one side has only one trig function, try to write the other side in terms of that one.

6. Factor.

7. Use the "prayer" method: Multiply top and bottom by the same term and pray that everything works out.

8. If nothing else works, change everything to sines and cosines.

NOTE

Some people do step 8 first. It does work many times. However, many times it makes the problem much longer since it always increases the number of terms.

ALSO NOTE

Please do not get discouraged. For almost everybody, these manipulative skills do not come easily. Only through lots of practice and much heartache does anyone become good.

Now let us give some examples. Remember—patience and practice.

EXAMPLE 21—

Add the fractions.

$$\frac{\tan x}{1 + \sec x} - \frac{\tan x}{1 - \sec x} = 2 \csc x$$

Multiply out top and bottom.

$$\frac{(\tan x)(1 - \sec x) - (\tan x)(1 + \sec x)}{(1 + \sec x)(1 - \sec x)} =$$

Simplify top; identity bottom.

$$\frac{\tan x - (\tan x)(\sec x) - \tan x - (\tan x)(\sec x)}{1 - \sec^2 x} =$$

Reduce; change to sines and cosines and finally reduce. Then note the last identity.

$$\frac{-2(\tan x)(\sec x)}{-\tan^2 x} =$$

$$\frac{2 \sec x}{\tan x} = \frac{2/\cos x}{\sin x/\cos x} = \frac{2}{\sin x} = 2 \csc x$$

Whew! Fortunately they are not all like this.

EXAMPLE 22—

Split into three fractions.

$$\frac{\cos x + \sin x - \sin^3 x}{\sin x} = \cot x + \cos^2 x$$

Identity and algebra.

$$\frac{\cos x}{\sin x} + \frac{\sin x}{\sin x} - \frac{\sin^3 x}{\sin x} =$$

$$\cot x + 1 - \sin^2 x = \cot x + \cos^2 x$$

Identity.

EXAMPLE 23—

$$\sin^4 x - \cos^4 x = \sin^2 x - \cos^2 x$$

Factor the left—more complicated.

$$(\sin^2 x - \cos^2 x)(\sin^2 x + \cos^2 x) = \sin^2 x - \cos^2 x$$

$\sin^2 x + \cos^2 x = 1$.

EXAMPLE 24—

$$\cos^4 x - 2\cos^2 x + 1 = \sin^4 x$$

$\cos^2 x = 1 - \sin^2 x$.

$$(1 - \sin^2 x)^2 - 2(1 - \sin^2 x) + 1 =$$

Hammer it out, simplifying.

$$1 - 2\sin^2 x + \sin^4 x - 2 + 2\sin^2 x + 1 = \sin^4 x$$

EXAMPLE 25—

$$\frac{\sin x}{1 + \cos x} = \frac{1 - \cos x}{\sin x}$$

Prayer method—multiply the top and bottom by 1 − cos x and pray it works.

$$\frac{(\sin x)(1 - \cos x)}{(1 + \cos x)(1 - \cos x)} =$$

Hint: **Do not multiply out the top because you want 1 − cos x left on the top.**

$$\frac{(\sin x)(1 - \cos x)}{1 - \cos^2 x} =$$

Identity.

$$\frac{(\sin x)(1 - \cos x)}{\sin^2 x} = \frac{1 - \cos x}{\sin x}$$

Well, that's a sampling. Some are easy (when you see them); some are hard; some are long; some are short. Try to have fun. Learn from your wrong turns. But don't quit. Practice, and you will get better and better!!!!!

TRIG EQUATIONS

Trig equations are solved the same way as algebraic equations, with two differences: Sometimes you use trig identities; and after you solve the equations, you still must solve for the angles.

Trig equations seem a lot harder than they should be because most books group all of them together instead of talking about each individual kind. What I've tried to do is to give you one of most of the types.

EXAMPLE 26—

$$2 \sin x - 3^{1/2} = 0$$

Let's start out slowly.

$$\sin x = \frac{3^{1/2}}{2}$$

Sine is positive in I and II. Draw the triangles!!

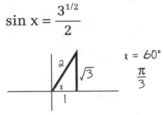

EXAMPLE 27—

$$2 \sin x \cos x - \cos x = 0$$

Take out a common factor.

$$(\cos x)(2 \sin x - 1) = 0$$

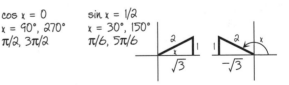
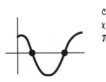

NOTE

When the sine or cosine = 0, +1, or −1, the angle is a multiple of 90°. The easiest way to find out which one is to draw y = sin x or y = cos x and look for the high points, low points, or intercepts.

EXAMPLE 28—

$$2 \sin^2 x - \sin x - 1 = 0$$

Factor the trinomial.

$$(2 \sin x + 1)(\sin x - 1) = 0$$

$\sin x = -\frac{1}{2}$

or $\sin x = 1$

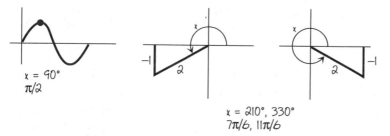

x = 90°
π/2

x = 210°, 330°
7π/6, 11π/6

EXAMPLE 29—

$3 \sin^2 x - 5 \sin x + 2 = 0$

$(3 \sin x - 2)(\sin x - 1) = 0.$ $\sin x = 1$; $\sin x = 2/3$ I, II.

As above, $\sin x = 1$ means $x = 90°$ or $\pi/2$. $\sin x = 2/3$, in the first two quadrants, can be found only by using the calculator. Most calculators work this way to calculate the answer: 2 divided by 3 = inverse, second, sin. Answers are 42° and 180° − 42° = 138°.

EXAMPLE 30—

$2 \sin^3 x - \sin x = 0$

A lot of answers. $(\sin x)(2 \sin^2 x - 1) = 0.$ $\sin x = 0$ or $\sin^2 x = 1/2$. So $\sin x = \pm 1/2^{1/2}$.

EXAMPLE 31—

$x = 0°, 180°, 45°, 135°, 225°, 315°,$ or $0, \pi, \pi/4, 3\pi/4, 5\pi/4, 7\pi/4$

$\cos^2 x + \cos x - 1 = 0.$ We must use the quadratic formula: $a = 1, b = 1, c = -1$:

$$\cos x = \frac{-b \pm (b^2 - 4ac)^{1/2}}{2a} = \frac{-1 \pm [1 - 4(1)(-1)]^{1/2}}{2(1)}$$

$$= \frac{-1 \pm 5^{1/2}}{2} = \frac{-1 \pm 2.24}{2}$$

One value is $(-1 - 2.24)/2 = -1.62$. No solution since the cosine is never less than -1. Another value is $(-1 + 2.24)/2 = 0.62$. Cosine is positive in the first and fourth quadrants. Using the calculator, we find $x = 52°$ and $360° - 52° = 308°$.

EXAMPLE 32—

$\sin^2 x + \cos x - 1 = 0$

Trig substitution: $\sin^2 x = 1 - \cos^2 x$:

$(1 - \cos^2 x) + \cos x - 1 = 0$ or $-\cos^2 x + \cos x = 0$

$-\cos x(\cos x - 1) = 0$ $\cos x = 0$ or 1

 $x = 0°, 90°, 270°,$ and $360°$ or
$0, \pi/2,$ and $3\pi/2$

EXAMPLE 33—

$\tan^2 x + \sec x - 1 = 0$

Trig substitution: $\tan^2 x = \sec^2 x - 1$:

$(\sec^2 x - 1) + \sec x - 1 = 0$ or $\sec^2 x + \sec x - 2 = 0$

$(\sec x + 2)(\sec x - 1) = 0$

$\sec x = -2$ or $\cos x = -\frac{1}{2}$ $\sec x = 1$ or $\cos x = 1$

$x = 120°, 240°,$ and $0°$
$2\pi/3, 4\pi/3,$ and 0

$x = 120°$ and $240°$ or $2\pi/3$ and $4\pi/3$ $x = 0°$ or 0

EXAMPLE 34—

$\sin x + \csc x = 2$

$\sin x + 1/\sin x = 2$ or $\sin^2 x + 1 = 2 \sin x$ or
$\sin^2 x - 2 \sin x + 1 = 0$ or $(\sin x - 1)^2 = 0$ or $\sin x = 1$.
So $x = 90°$ or $\pi/2$.

Let's wind up with a slightly messy one.

EXAMPLE 35—

$\sin x - 3^{1/2} \cos x = 1$

$\sin x - 1 = 3^{1/2} \cos x$

$(\sin x - 1)^2 = (3^{1/2} \cos x)^2$

$\sin^2 x - 2 \sin x + 1 = 3 \cos^2 x$ or $\sin^2 x - 2 \sin x + 1 =$
$$3(1 - \sin^2 x)$$

Isolate one of the trig functions, the messier one, the one with the square root of 3, and square both sides.

or $4 \sin^2 x - 2 \sin x - 2 = 0$ or $2(2 \sin x + 1)(\sin x - 1) = 0$

$\sin x = 1$ or $-\frac{1}{2}$. $x = 90°$ or $210°$ or $330°$

Since we squared both sides, we may have introduced extraneous solutions or nonsolutions. (This may also be the case if we start with a trig function in the denominator.) We must substitute the answers into the original equation:

$x = 90°$. $\sin 90° - 3^{1/2} \cos 90° = 1$. $1 - 0 = 1$. $90°$ is OK.

$x = 210°$. $\sin 210° - 3^{1/2} \cos 210° = 1$.
$$-\frac{1}{2} - 3^{1/2}(-3^{1/2}/2) = 1. \ -\frac{1}{2} + 3/2 = 1. \ 210° \text{ is OK.}$$

$x = 330°$. $\sin 330° - 3^{1/2} \cos 330° = 1$.
$$-\frac{1}{2} - 3^{1/2}(+3^{1/2}/2) = 1.$$

$-\frac{1}{2} - 3/2 \neq 1$. $330°$ is extraneous.

The answers are $90°$ and $210°$ or $\pi/2$ and $7\pi/6$.

DOMAINS AND RANGES OF THE TRIG FUNCTIONS

Let's list the ranges and domains of the trig functions. Then we will discuss them.

	Domain	Range
$y = \sin x$	All real numbers	$-1 \le y \le 1$
$y = \cos x$	All real numbers	$-1 \le y \le 1$
$y = \tan x$	All real numbers except $90° \pm n180°$, n integer	All real numbers
$y = \cot x$	All real numbers except $0° \pm n180°$	All real numbers
$y = \sec x$	Same as tan x	All real numbers except $-1 < y < 1$
$y = \csc x$	Same as cot x	Same as sec x

From the curve sketch, the domain and range for the sin x and cos x should be seen. tan x = y/x and sec x = r/x are defined everywhere except where the bottom, x, is 0. That occurs on the y axis. The angles are 90°, 270°, −90°, etc.

A similar argument exists for the domain of the cot x and csc x except the x axis is no good. Angles are 0°, 180°, −180°, etc. Range for the tangent is seen from the graph. The cotangent, the reciprocal of the tangent, is the same. The range for the sec x and csc x is the reciprocals of the sin x and cos x. However, all can be +1 or −1.

It is of value to know this chart, especially sin x, cos x, and tan x.

DOUBLE ANGLES, HALF ANGLES, SUM AND DIFFERENCE OF TWO ANGLES

In the past, I would have just listed the following formulas. But very recently a student showed me it is necessary to derive at least the first formula for belief. Believing a formula is true somehow makes it easier to learn. I don't know why, but it does to many students.

We will derive the cos (A − B). We need to use the unit circle, that is, the circle $x^2 + y^2 = 1$, the circle center at the origin, r = 1. On this circle, cos A = x/r, but r = 1. So x = cos A. sin A = y/r = y/1 = y. So y = sin A. So every point (x, y) on the unit circle is given by (cos A, sin A), as in the first figure here. Aren't mathematicians clever?!

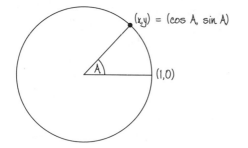

On the second circle, we draw the angles A, B, and A − B. We label the points on the circle as we did in the first picture. We draw the chords. We learned in geometry that in a circle, if there are equal central angles (both A − B), there are equal chords. Chord lengths mean the distance formula. We will use the square of the distance formulas so that there are no square roots.

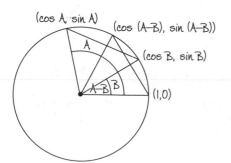

$$(x_2 - x_1)^2 + (y_2 - y_1)^2 = (x_4 - x_3)^2 + (y_4 - y_3)^2$$

$$[\cos (A - B) - 1]^2 + [\sin (A - B) - 0]^2$$
$$= (\cos A - \cos B)^2 + (\sin A - \sin B)^2$$

We just multiply it out, and that will be the answer.

$$\sin^2 (A - B) + \cos^2 (A - B) - 2 \cos (A - B) + 1$$
$$= \sin^2 A + \cos^2 A + \sin^2 B + \cos^2 B$$
$$- 2(\cos A)(\cos B) - 2(\sin A)(\sin B)$$

sin² + cos² of three different angles is 1.

$$1 - 2 \cos (A - B) + 1$$
$$= 1 + 1 - 2(\cos A)(\cos B) - 2(\sin A) (\sin B)$$

The reason? 1 + 1 = 2. I love to say that in an advanced class. Now subtract 2 from both sides and divide everything by –2.

$$2 - 2 \cos (A - B) = 2 - 2(\cos A)(\cos B) - 2(\sin A)(\sin B)$$

$$\cos (A - B) = \cos A \cos B + \sin A \sin B$$

I hope you will trust me for the rest of the formulas. They are derived in virtually every trig book. There are two reasons I derived this one: Too many people think that to find the cos (A – B) you use the distributive law; and some others can't imagine why a formula like cos (A – B) requires knowing the sines as well as the cosines. I hope you are convinced.

Here is the second list of identities. In order to be very good, you should memorize this list, even if your instructor tells you you do not have to.

$$\cos (A - B) = \cos A \cos B + \sin A \sin B$$
$$\cos (A + B) = \cos A \cos B - \sin A \sin B$$

$$\sin (A - B) = \sin A \cos B - \cos A \sin B$$
$$\sin (A + B) = \sin A \cos B + \cos A \sin B$$

$$\tan (A - B) = \frac{\tan A - \tan B}{1 + \tan A \tan B}$$

$$\tan (A + B) = \frac{\tan A + \tan B}{1 - \tan A \tan B}$$

Half angle: Precalc, $\cos (A/2) = \pm[(1 + \cos A)/2]^{1/2}$

Calc, $\cos^2 A = (1 + \cos 2A)/2$

Half angle: Precalc, $\sin (A/2) = \pm[(1 - \cos A)/2]^{1/2}$

Calc, $\sin^2 A = (1 - \cos 2A)/2$

Half angle: $\tan (A/2) = \pm[(1 - \cos A)/(1 + \cos A)]^{1/2}$

$= (1 - \cos A)/\sin A = \sin A/(1 + \cos A)$

$\cos 2A = \cos^2 A - \sin^2 A = 2 \cos^2 A - 1 = 1 - 2 \sin^2 A$

$\sin 2A = 2 \sin A \cos A$

$$\tan 2A = \frac{2 \tan A}{1 - \tan^2 A}$$

This third list is a list of identities some of which are nice to know and some of which are not done in every class:

$\sin (90° - A) = \cos A$ $\cos (90° - A) = \sin A$ $\sin (180° - A) = \sin A$

$\sin (-B) = -\sin B$ $\cos (-B) = \cos B$ $\tan (-B) = -\tan B$

$2 \sin A \cos B = \sin (A + B) + \sin (A - B)$

$2 \cos A \sin B = \sin (A + B) - \sin (A - B)$

$2 \cos A \cos B = \cos (A + B) + \cos (A - B)$

$2 \sin A \sin B = -\cos (A + B) + \cos (A - B)$

$\sin C + \sin D = 2 \sin \frac{1}{2} (C + D) \cos \frac{1}{2} (C - D)$

$\sin C - \sin D = 2 \cos \frac{1}{2} (C + D) \sin \frac{1}{2} (C - D)$

$\cos C + \cos D = 2 \cos \frac{1}{2} (C + D) \cos \frac{1}{2} (C - D)$

$\cos C - \cos D = -2 \sin \frac{1}{2} (C + D) \sin \frac{1}{2} (C - D)$

In most courses, the majority of the problems are numerical. There are usually a few identities and a few equations. That is what we will do.

EXAMPLE 36—

Given cos A = 4/5 in quadrant IV and sin B = −5/13 in III. Find sin 2B, cos 2A, tan (A − B), sin (A/2), and cos (A/2).

First draw the appropriate triangles. Note the 3-4-5 and 5-12-13 Pythagorean triples.

sin A = −3/5, cos A = 4/5, tan A = −3/4. sin B = −5/13, cos B = −12/13, tan B = 5/12.

    The first three, you just plug into the appropriate formula:

A. $\sin 2B = 2 \sin B \cos B = 2\left(\dfrac{-5}{13}\right)\left(\dfrac{-12}{13}\right) = \dfrac{120}{169}$

B. Any of the three formulas is OK.

$$\cos 2A = \cos^2 A - \sin^2 A = \left(\dfrac{4}{5}\right)^2 - \left(\dfrac{-3}{5}\right)^2 = \dfrac{7}{25}.$$

C. $\tan (A - B) = \dfrac{\tan A - \tan B}{1 + \tan A \tan B}$

$$= \dfrac{\left(\dfrac{-3}{4}\right) - \left(\dfrac{5}{12}\right)}{1 + \left(\dfrac{-3}{4}\right)\left(\dfrac{5}{12}\right)} = \dfrac{-56}{33}$$

For sin A/2 and cos A/2, we use the precalc formulas, but there is a little more since we have the ± sign. We would like to know which sign. Since A is in IV,

270° < A < 360°. So 135° < A/2 < 180°, dividing every-thing by 2. A/2 is in quadrant II. sin (A/2) is positive; cos (A/2) is negative.

$$\sin\left(\frac{A}{2}\right) = \left[\frac{(1-\cos A)}{2}\right]^{1/2} = \left[\frac{\left(1-\frac{4}{5}\right)}{2}\right]^{1/2}$$

$$= \left(\frac{1}{10}\right)^{1/2} = \frac{10^{1/2}}{10}$$

$$\cos\left(\frac{A}{2}\right) = -\left[\frac{(1+\cos A)}{2}\right]^{1/2} = -\left[\frac{\left(1+\frac{4}{5}\right)}{2}\right]^{1/2}$$

$$= -\left(\frac{9}{10}\right)^{1/2} = \frac{-3(10)^{1/2}}{10}$$

EXAMPLE 37—

Use three different formulas to calculate sin 75°.

A. $\sin(75°) = \sin(45° + 30°) = \sin 45° \cos 30°$

$$+ \cos 45° \sin 30° = \left(\frac{2^{1/2}}{2}\right)\left(\frac{3^{1/2}}{2}\right) + \left(\frac{2^{1/2}}{2}\right)\left(\frac{1}{2}\right)$$

$$= \frac{(6^{1/2} + 2^{1/2})}{4}$$

B. $\sin(75°) = \sin(120° - 45°)$

$$= \sin 120° \cos 45° - \cos 120° \sin 45°$$

$$= \left(\frac{3^{1/2}}{2}\right)\left(\frac{2^{1/2}}{2}\right) - \left(-\frac{1}{2}\right)\left(\frac{2^{1/2}}{2}\right) = \frac{(6^{1/2} + 2^{1/2})}{4}$$

C. $\sin(75°) = \sin \frac{1}{2}(150°) = \left[\frac{(1-\cos 150°)}{2}\right]^{1/2}$

$$= \frac{\left[1 - \left(\dfrac{3^{1/2}}{2}\right)\right]^{1/2}}{2^{1/2}} = \frac{\left[1 + \left(\dfrac{3^{1/2}}{2}\right)\right]^{1/2}}{2^{1/2}}$$

You can use a calculator to check to see that these answers are all the same. It is a very difficult exercise to show that the answers are exactly the same. Try it if you are brave. Here's a nasty problem.

EXAMPLE 38—

Find sin A if tan 2A = −24/7 in quadrant II.

If 2A is in II, A is in I:

Cross multiply—
−24(1 − tan² A) =
14 tan A.

$$\tan 2A = \frac{2 \tan A}{1 - \tan^2 A} = \frac{-24}{7}$$

Rearranging, we get 24 tan² A − 14 tan A − 24 = 0 or 2(3 tan A − 4)(4 tan A + 3) = 0. tan A = −3/4— extraneous since A is in I. tan A = 4/3. So sin A = 4/5. (A 3-4-5 triangle.)

EXAMPLE 39—

Show sin (180° − A) = sin A.

This is just a hammer-it-out identity: sin (180° − A) = (sin 180°)(cos A) − (cos 180°)(sin A) = sin A since sin 180° = 0 and cos 180° = −1.

EXAMPLE 40—

sin (A + B) − sin (A − B) = 2 cos A sin B

Another hammer-it-out:

sin (A + B) − sin (A − B) = sin A cos B + cos A sin B

$$- (\sin A)(\cos B) + (\cos A)(\sin B)$$

$$= 2 \cos A \sin B$$

EXAMPLE 41—

Show sin 3A = 3 sin A − 4 sin³ A.

This is a little more work:

sin 3A = sin (2A + A) = sin 2A cos A + cos 2A sin A

$$= 2(\sin A)(\cos A)(\cos A) + (1 - 2 \sin^2 A)(\sin A)$$

$$= 2 \sin (1 - \sin^2 A) + (1 - 2 \sin^2 A)(\sin A)$$

$$= 3 \sin A - 4 \sin^3 A$$

In some places you have to know this one too!!!!!
   In doing equations involving double or triple angles, there are two basic types: one where the double (triple) angle goes and one where it stays.

EXAMPLE 42—

cos 2x + sin x = 0

Trig substitute for cos 2x, the one involving sin x:

(1 − 2 sin² x) + sin x = 0

or   −2 sin² x + sin x + 1 = 0

or   −1(2 sin x + 1)(sin x − 1) = 0

We've done these many times before (the more you do, the better you get). sin x = −½ or 1. So x = 210°, 330°, and 90° or 7π/6, 11π/6, and π/2. (If you forget, always draw the triangles!!!!!)
   The other type (there could be combinations of this with the last type) is where the multiple angle stays. The techniques are the same as in the first section on identities. So we'll just do a basic one.

EXAMPLE 43—

sin (3A) = ½

If 0° ≤ A ≤ 360°, then 0° ≤ 3A ≤ 1080°.

Since the sine is periodic, period 360°, we must add 360° to each 3A angle as long as the total is less than 1080°, so that when we divide by 3, all the angles are between 0° and 360°. 3A = 30°, 150°, 30° + 360°, 150° + 360°, 30° + 720°, and 150° + 720°. 3A = 30°, 150°, 390°, 510°, 750°, and 870°. So A = 10°, 50°, 130°, 170°, 250°, and 290°.

NOTE 1
If you add 360° too many times, when you divide by 3 the answer will be more than 360°.

NOTE 2
If our example were sin 9A, you'd add 360° eight times to each angle in the answer.

INVERSE TRIG FUNCTIONS

One of the things that amazes me is how often inverse trig functions are taught apart from understanding inverse functions. If you have forgotten, review inverse functions, which are in Chapter 8.

If you recall, there must be a 1:1 correspondence between domain and range—usually indicated by a function that increases only or decreases only. As you see, y = sin x does not fit this description.

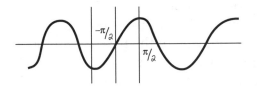

We have to restrict the original domain, as pictured. So the only angles we consider are $-\pi/2 \leq x \leq \pi/2$. Remember this!!!!!!!

DEFINITION

Arcsin x or inverse sin (notation Arcsin x or $\sin^{-1}$ x)
Domain $-1 \le x \le 1$ range $-\pi/2 \le y \le \pi/2$
Define Arcsin x = y if originally sin y = x.

EXAMPLE 44—

Find the arcsin ½.

Read, "We are looking for the angle whose sine is ½".
   Since the arc sine is defined in quadrants I and IV
only—first, positive angles, and fourth, negative
angles—*the only answer is* $\pi/6$ since sin ($\pi/6$) = ½ and
$\pi/6$ is in quadrant I.

EXAMPLE 45—

$$\sin^{-1}\left(\frac{-1}{2^{1/2}}\right)$$

The answer must be in quadrant IV and a negative
angle. Answer: $-\pi/4$.
   In most schools one discusses arcsin, arc cos, and
arctan. We will too. You should learn this chart!!!!!

Trig Function	Domain	Range
$\sin^{-1}$	$-1 \le x \le 1$	$-\pi/2 \le y \le \pi/2$
$\cos^{-1}$	$-1 \le x \le 1$	$0 \le y \le \pi$
$\tan^{-1}$	$-\infty < x < \infty$	$-\pi/2 < y < \pi/2$

EXAMPLE 46—

   A. $\tan^{-1} 1$

   B. $\tan^{-1} (-1/3^{1/2})$

$\angle A = \pi/4$ $\angle B = -\pi/6$

EXAMPLE 47—

A. $\cos^{-1}(\frac{1}{2})$

B. $\cos^{-1}(-3^{1/2}/2)$

Remember: arcsin and arctan are in quadrants I and IV; arc cos is in I and II.

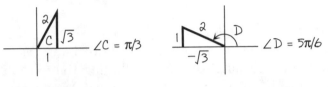

EXAMPLE 48—

sec (Arctan 5/7)

We are looking for the secant of the angle whose tangent is 5/7.

Draw a triangle whose tangent is 5/7 in quadrant I. Find the third side and then get the secant, which is r/x. $r = (7^2 + 5^2)^{1/2} = 74^{1/2}$. The secant is $74^{1/2}/7$.

EXAMPLE 49—

cot $(\sin^{-1} v)$

We are looking for the cotangent of the angle whose sine is $v = v/1$.

Draw the triangle. $\sin = y/r = v/1$. $x = (1 - v^2)^{1/2}$. cot is $x/y = (1 - v^2)^{1/2}/v$. This is used in calc. It is not hard with a little practice. In fact, it is more fun with letters.

EXAMPLE 50—

$\sin (\cos^{-1} p + \sin^{-1} q)$

Let $A = \cos^{-1} p$ and $B = \sin^{-1} q$. So $\cos A = p$ and $\sin B = q$.

Draw both triangles. We are looking for $\sin (A + B) = (\sin A)(\cos B) + (\cos A)(\sin B) = (1 - p^2)^{1/2}(1 - q^2)^{1/2} + pq$.

I like these! Hope you will too.

Right-Angle Trig

If the triangle is in quadrant I, we can take it out of the x-y plane and look at the triangle a different way:

$\sin A = y/r = \text{opposite/hypotenuse}$ $\cos A = x/r = \text{adjacent/hypotenuse}$

$\tan A = y/x = \text{opposite/adjacent}$ $\cot A = x/y = \text{adjacent/opposite}$

$\sec A = r/x = \text{hypotenuse/adjacent}$ $\csc A = r/y = \text{hypotenuse/opposite}$

We could do many, many examples. However, most are of the same general type. So we will settle for two problems: a relatively simple one and a relatively complicated one.

In some problems we have the *angle of elevation* (the angle you look up at) or the *angle of depression* (the angle you look down at from above). As pictured here, they are equal:

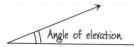

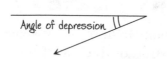

EXAMPLE 51

A man is looking down from the top of a 120-foot lighthouse. The angle of depression to a boat is 20°. How far away is the boat?

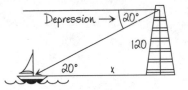

Try to get the variable in the numerator. cot 20° = x/120. So x = 120 cot 20°. By calculator, the answer is 330 feet.

EXAMPLE 52

John is 200 feet from a building on which is located a tall antenna. The angle of elevation to the bottom of the antenna is 70°. The angle of elevation to the top of the antenna is 78°. How tall is the antenna?

There are two unknowns: the height of the building and the height of the antenna. We always want to try to get the unknown we don't want (the building) in the numerator. It makes the problem easier. In this problem, both unknowns are in the numerator when you use tangent.

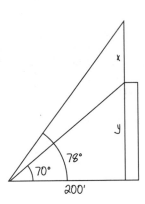

$$\tan 78° = \frac{(x + y)}{200} \qquad x + y = 200 \tan 78°$$

Subtract; factor.

$$\tan 70° = \frac{y}{200} \qquad y = 200 \tan 70°$$

$$x = 200(\tan 78° - \tan 70°) = 394 \text{ feet (big!)}$$

THE LAW OF SINES

When we don't have a right angle, we must use the law of sines or the law of cosines.

DEFINITION

Law of sines: a/sin A = b/sin B = c/sin C.

NOTE

For this exercise we will say sin 45° = 0.7, which is approximately correct.

We use the law of sines whenever we have any two angles and a side or two sides and an angle opposite one of those sides.

NOTE

Side is small letter; angle opposite is the same letter, only capitalized.

Two Angles and a Side

If we have two angles, we have the third angle. With a side, we learned, in geometry, that triangles are congruent by angle, side, angle. Therefore, one solution is possible.

EXAMPLE 53—

A = 56°, B = 73°, and a = 20. Find all the other parts.

C = 180° − (56° + 73°) = 51°. a/sin A = b/sin B.
20/sin 56° = b/sin 73°. b = 20 sin 73°/sin 56°. b = 263.
a/sin A = c/sin C. 20/sin 56° = c/sin 51°. c = 20 sin 51°/
sin 56°. c = 19.

Two Sides and an Angle Opposite One of Those Sides

If you draw the triangle, we have side, side, angle. From geometry, we know triangles are *not* congruent. This is called the *ambiguous case.* No, one, or two triangles are possible. Let us give one example of each.

EXAMPLE 54—

Let a = 10, A = 30°, b = 50.

a/sin A = b/sin B. 10/sin 30 = 50/sin B. sin B = (50 sin 30)/10 = 2.5. No triangle since the sine is never bigger than 1.

EXAMPLE 55—

A = 135°, a = 70, b = 50.

a/sin A = b/sin B. 70/sin 135° = 50/sin B. sin B = (50 sin 135)/70. sin B = ½. B = 30° or 180° − 30° = 150°.
  135° + 30° = 165° is OK. 135° + 150° = 285° is no good—the sum of angles of a triangle is 180°. One triangle is possible. A = 135°, B = 30°, so C = 15°. a = 70, b = 50. a/sin A = c/sin C. 70/sin 135° = c/sin 15°. c = 70 sin 15°/sin 135°. c = 26.

EXAMPLE 56—

C = 30°. c = 5. d = 7.

c/sin C = d/sin D. 5/sin 30 = 7/sin D. sin D = (7 sin 30°)/5 = 0.7. D = 45° or D′ = 180° − 45° = 135°. 30 + 45, OK. 30 + 135, OK. Two triangles!!!

Triangle 1: C = 30°, D = 45°, E = 105°. c = 5, d = 7, e/sin 105° = 5/sin 30°. e = 9.7.

Triangle 2: C = 30°, D′ = 135°, E′ = 15°. c = 5, d = 7, e′/sin 15° = 5/sin 30°. e′ = 2.6.

EXAMPLE 57—

The angle between Zeb and Sam as seen by Don is 70°. The angle between Don and Zeb as seen by Sam is 62°. Zeb and Sam are 70 feet apart. How far apart are Don and Sam?

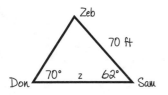

Angle at Zeb is 48°. z/sin 48° = 70/sin 70°. z = 55 feet.

THE LAW OF COSINES

We use the law of cosines whenever we have three sides or two sides and the included angle.

DEFINITION

Law of cosines: $c^2 = a^2 + b^2 - 2ab \cos C$. Again, note angle C opposite side c. In each case only one triangle is possible, since triangles are congruent by side, side, side or side, angle, side.

EXAMPLE 58—

$a = 3, b = 5, c = 7$

$c^2 = a^2 + b^2 - 2ab \cos C$. $7^2 = 3^2 + 5^2 - 2(3)(5) \cos C$. Careful of the arithmetic. $15 = -30 \cos C$. $\cos C = -\frac{1}{2}$. $C = 120°$ (quadrant II).

NOTE

Whenever you have four parts of a triangle, it is easier to use the law of sines. $3/\sin A = 7/\sin 120°$. $A = 22°$. $B = 180° - (22 + 120) = 38°$.

EXAMPLE 59—

$x = 10, y = 20, Z = 40°$

$z^2 = x^2 + y^2 - 2xy \cos Z$. $z^2 = 10^2 + 20^2 - 2(10)(20)$ cos 40°. $z = 13.9$. $13.9/\sin 40° - 10/\sin X$. $\sin X = 10(\sin 40°)/13.9$. $X = \sin^{-1}(10 \sin 40°/13.9)$. $X = 28°$. $Y = 180° - (28° + 40°) = 112°$.

EXAMPLE 60—

A plane travels east for 200 miles. It turns at a 25° angle to the north for 130 miles. How many miles from home is the plane?

$$x^2 = 200^2 + 130^2 - 2(200)(130) \cos 155°$$

$$x = 323 \text{ miles}$$

I hope this chapter is sufficient to give you a thorough knowledge of trig. If you do get this knowledge, it will more than prepare you for calculus.

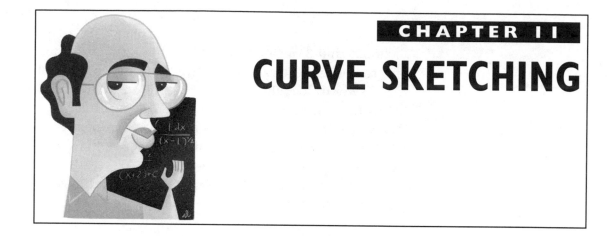

CHAPTER 11
CURVE SKETCHING

As I've said in my other books, this is the topic I teach better than anyone else in the world. After reading this chapter, you will be nearly as good. This chapter has been written in a slightly different order than the discussion of curve sketching in the calc book. Of course the part that requires calculus has been omitted here.

POLYNOMIALS

We will sketch polynomials looking only at intercepts and the exponents of x intercepts. With just a little practice, you will become lightning fast. First let's look at the individual cases.

EXAMPLE 1—

$y = 6(x - 3)^8$

We want to look at this around its x intercept. We ignore the exponent and see that the intercept is (3, 0). We would like to see it a little to the right of 3, at 3^+, and to the left of 3, at 3^-. Remember, 3^+ means

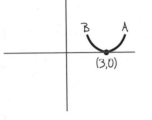

3.000001 and 3^- means 2.999999. When we substitute, we are not interested in the answer but only the sign of the answer to see whether a nearby point is above or below the x axis. Point A: $x = 3^+$. $3^+ - 3$ is positive. Positive to the eighth power is positive. Six times a positive number is positive. Point A is above the x axis. Let B have $x = 3^-$. $3^- - 3$ is negative. A negative to the eighth power (even power) is *positive*. Six times a positive is positive. B is also above the x axis.

EXAMPLE 2—

$$y = -5(x - 4)^{88}$$

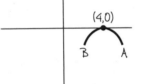

Intercept (4, 0). A: $x = 4^+$. $4^+ - 4$ is positive. Positive to the eighty-eighth power is positive; -5 times a positive is negative. Point A is below the x axis. For B, $x = 4^-$. $4^- - 4$ is negative. Negative to the eighty-eighth power is positive; -5 times a positive is again negative.

   To summarize, if you have an even power, the curve does *not* cross at the intercept.

EXAMPLE 3—

$$y = 9(x - 6)^5$$

Intercept (6, 0). A: $x = 6^+$. y is positive. A is above the x axis. B: $x = 6^-$. $6^- - 6$ is negative. Negative to the fifth power (odd) is negative; 9 times a negative is negative. B is below the x axis.

EXAMPLE 4—

$$y = -8(x + 4)^7$$

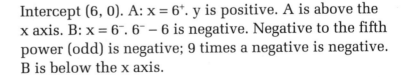

Intercept (−4, 0). A: $x = (-4)^+$. $(-4)^+ = -3.999999999$. $(-4)^+ + 4$ is positive. Positive to the seventh power is positive. Positive times (−8) is negative. (That is,

point A.) Point A is below the x axis. B, x = (−4)⁻.
(−4)⁻ + 4 is negative. Negative to the seventh power is
negative; −8 times a negative is positive. B is above
the x axis.

In summary, if the exponent is odd, there is a crossing at the x intercept.

NOTE

If y = 4(x − 3), the exponent of (x − 3) is 1, which is
odd. So there would be a crossing at (3, 0).

EXAMPLE 5—

$$y = (x - 1)^2(x - 2)(x - 5)$$

x intercepts whenever y = 0. (1, 0), (2, 0), (5, 0). y intercept
at x = 0. We must substitute x = 0. y = $(0 - 1)^2(0 - 2)$
$(0 - 5)$ = 10. (0, 10). We must substitute one number to
tell us where the graph starts. That number is to the
right of the rightmost intercept. In this case x = 5.1.
$(5.1 - 1)^2(5.1 - 2)(5.1 - 5)$ is positive, so the point is
above the x axis. Remember, we care about the sign
only. The power of (x − 5) is 1, which is odd. So it
crosses at (5, 0), and the point is below the x axis when
we near (2, 0). The power of (x − 2) is odd, and the
crossing is under to above as we near (1, 0). The $(x - 1)^2$
term has power 2, which is even. The graph is above to
the right. So it is above to the left. The curve goes
through (0, 10). Both ends head to plus infinity since
there are no other x intercepts, and to the right of (5, 0)
and to the left of (1, 0) the graph is above the x axis.
The sketch looks as shown here.

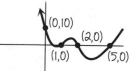

With a little practice, you will see how quick this
can get. Let's give a very easy example.

EXAMPLE 6—

$$y = x^3 - 2x^2 - 8x$$

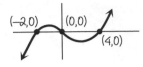

Factor $y = x(x - 4)(x + 2)$. Intercepts are $(0, 0)$, $(4, 0)$, and $(-2, 0)$. Substitute $x = 4.1$ in factored form. You will see that y is positive. To the right of $(x - 4)$, the sketch is above the x axis. All the exponents are odd. So the sketch goes cross, cross, cross.

NOTE

If we have the intercept $(0, 0)$, it is an x intercept, but it is the only y intercept. We do not have to get the y intercept separately if we have $(0, 0)$.

ALSO NOTE

Don't forget $(0, 0)$ if it is there.

ALSO NOTE

See how fast this goes: cross, cross, cross.

EXAMPLE 7—

$y = (x - 2)^2(3 - x)$

$(2, 0)$, $(3, 0)$ (careful of this one). $x = 0$, $y = (-2)^2(3) = 12$. $(0, 12)$. $x = 3.1$ so $y = (3.1 - 2)^2(3 - 3.1)$ is negative. So the curve starts below the x axis, with the right end going to minus infinity. The power of $(3 - x)$ is odd. So there is a crossing from below to above. At $(2, 0)$, the power is even—$(x - 2)^2$—so the graph stays above the x axis at $(2, 0)$, goes through $(0, 12)$, and heads to plus infinity. It looks as shown here.

EXAMPLE 8—

Let's take a mean-looking one and show how really easy it is: $y = x^3(x - 2)^4(x - 4)^5(x - 6)^6(x - 8)^7$.

Intercepts are $(0, 0)$, $(2, 0)$, $(4, 0)$, $(6, 0)$, and $(8, 0)$. Substitute $x = 8.1$. y is positive. To the right of $(8, 0)$ above. The right end goes to plus infinity.

$(x - 8)^7$, power is odd. The crossing is above to below.

$(x - 6)^6$, power is even. There is no crossing. Graph stays below.

$(x - 4)^5$, power is odd. The crossing is below to above. Remember, we always go from right to left.

$(x - 2)^4$, power is even. There is no crossing. Sketch stays above the x axis.

x^3, power is odd. The crossing is above to below, and the right end goes to minus infinity.

Its sketch is as shown here:

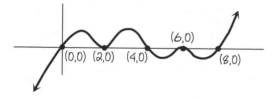

With a little practice you *will* do these in no time at all.

RATIONAL FUNCTIONS

We are next going to sketch rational functions—which are polynomials divided by polynomials. The intercept part is exactly the same. We need to introduce the concept of *asymptote.* For our purposes, an asymptote is a line to which a graph gets very close at the end but never hits.

You may have been told that a curve cannot hit an asymptote. That is not true. An asymptote is a straight-line approximation when either x or y goes to plus or minus infinity. In the middle of the graph, the curve is not a straight line. So the graph may hit the asymptote a number of times. At the end, though, the curve cannot hit it. Shown here is an example:

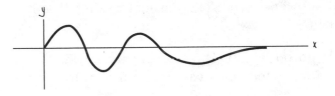

The x axis is the asymptote; the axis is hit four times, but not at the end.

Recall that the *degree* of a polynomial in one unknown is its highest exponent. The degree of y = $7x^4 - 9x^5 + 3x + 2$ is 5.

Vertical Asymptotes

The vertical asymptotes are lines. They are derived by setting the bottom of the fraction equal to 0.

NOTE 1

Polynomials have no denominators. Therefore, they have no asymptotes of any kind.

NOTE 2

The discussion of vertical asymptotes is very similar to that of intercepts.

EXAMPLE 9—

$$y = \frac{4}{(x - 6)^8}$$

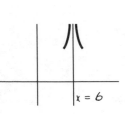

Vertical asymptote at x = 6. If we substitute 6^+, 6^- as before, y is positive. So on both sides of x = 6, the curve goes to plus infinity. The picture would look as shown here.

EXAMPLE 10—

$$y = \frac{-3}{(x-4)^{22}}$$

The vertical asymptote is at $x = 4$. When you substitute 4^+, 4^-—you can use 4.1 and 3.9—the value of y is negative. Both ends go to minus infinity. Again the curve looks as shown here.

To summarize, if a term has an even exponent in the *bottom,* both ends of the curve either go to plus infinity or go to minus infinity.

EXAMPLE 11—

$y = 7/(x-9)^5$

The vertical asymptote is at $x = 9$. If we substitute 9.1, y is positive, and the right side of $x = 9$ goes to plus infinity. If we substitute $x = 8.9$, the value of y is negative. The left side goes to minus infinity. The picture is as shown here.

EXAMPLE 12—

$y = -8/(x-3)$

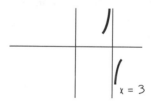

The vertical asymptote is at $x = 3$. Remember $x = 3$, power is 1, which is odd. $x = 3.1$, y is negative. To the right the curve goes to minus infinity (down). $x = 2.9$, y is positive. To the left the curve goes to plus infinity (up). The picture is again as shown here.

To summarize, if we have an odd exponent in the bottom, one end goes to plus infinity (up) and the other end (on the other side of the asymptote) goes to minus infinity (down).

Horizontal Asymptotes

There are two types of horizontal asymptotes.

Suppose $y = P(x)/Q(x)$. The degree of the top is less than the degree of the bottom. We always get the horizontal asymptote $y = 0$, the x axis. Let us verify this for one example.

The degree of the top is 2; the degree of the bottom is 3. The degree of the bottom is bigger. Horizontal asymptote is $y = 0$. Let us verify. Divide top and bottom by x^3. This is OK since x goes to plus or minus infinity and not to 0. We get . . .

EXAMPLE 13—

$$y = \frac{5x^2 + 8x}{4 + 9x^3}$$

$5/x$, $8/x^2$, and $4/x^3$ all go to 0 as x goes to plus or minus infinity. So $y \to 0/9$ $= 0$, as promised. In the future, do not do the work. If the bottom degree is bigger, $y = 0$ will always be the horizontal asymptote.

$$y = \frac{5/x + 8/x^2}{4/x^3 + 9}$$

Suppose the degree of the top is the same as the bottom; then the horizontal asymptote is

$$y = \frac{a}{b}$$
a is the coefficient of the highest power on top.

b is the coefficient of the highest power on the bottom.

EXAMPLE 14—

Degree of top and bottom is 3. The asymptote is $y = 5/(-9)$. Again let us verify, so that you believe. Divide by x^3.

$$y = \frac{5x^3 + 7x^2}{4 - 9x^3}$$

$$y = \frac{5 + 7/x}{4/x^3 - 9}$$

7/x and 4/x³ go to 0 as x goes to plus or minus infinity. So y → −(5/9).

An oblique asymptote occurs when the degree of the top is 1 more than the bottom. In order to get it, we long divide the bottom into the top. Groan—"Aarrgh!"

EXAMPLE 15—

$$y = \frac{x^2 - 4x + 4}{x - 1}$$

$$
\begin{array}{r}
x - 3 \;+ 1/(x-1) \\
x - 1 \overline{)x^2 - 4x + 4} \\
\underline{x^2 - x} \\
-3x + 4 \\
\underline{-3x + 3} \\
1
\end{array}
$$

The remainder $1/(x - 1)$ goes to 0 as x goes to plus or minus infinity. The oblique (slanted) asymptote is $y = x - 3$.

NOTE

If the degree of the top is more than 1 more than the bottom, there are no horizontal or oblique asymptotes.

ALSO NOTE

There is at most one horizontal or one oblique asymptote when you sketch rational functions. There might be none. There cannot be one of each.

EXAMPLE 16—

$$y = \frac{(x - 1)^4(x + 2)^3}{(x + 1)^8(x - 2)^1}$$

Now don't panic. Pretty soon you will be able to do this in less than two minutes. We do this very systematically:

x intercept: y = 0, top of the fraction = 0. (1, 0) and (–2, 0).

y intercept: x = 0, y = $(-1)^4(2)^3/1^8(-2)^1 = -4$. (0, –4).

Vertical asymptotes: Bottom = 0. x = –1, x = 2.

Horizontal asymptote: The degree of the top is 7 (if we multiplied out the top, and we wouldn't really do it on the grounds of sanity, the highest power is x^7, the only term we are interested in); the degree of the bottom is 9. Since the degree of the bottom is bigger, we get y = 0, the x axis.

We start the sketch. We always start from the rightmost x intercept or rightmost vertical asymptote. In this case we start at x = 2, the vertical asymptote. We substitute x = 2.1 and find that y will be a positive number. So the curve goes to plus infinity on the right side of x = 2:

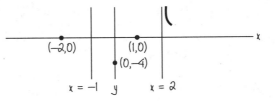

To continue the sketch, the power of (x – 2) is 1, an odd number. Since the right side went to plus infinity, the left side must go to minus infinity. We come up to the point (1, 0). $(x - 1)^4$ is an even exponent. So there is no crossing. The curve stays below, goes through (0, –4), and heads to minus infinity at x = –1:

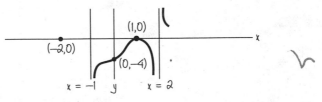

At $(x + 1)^8$, the power is even. Both ends are in the same location. In this case both ends are at minus infinity. The left side heads up to $(-2, 0)$. $(x + 2)^3$ has an odd power. So there is a crossing. Both ends head toward the horizontal asymptote, $y = 0$, the x axis. *Don't forget the ends!!!!!*

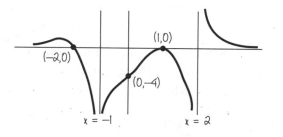

That's it. Again, do not panic. You will get it easily if you go over this example and the next few several times.

EXAMPLE 17—

$$y = \frac{2x^3}{(x - 3)^2(x - 1)} \qquad y = \frac{2x^3}{x^3 + \cdots}$$

x intercept: $y = 0$, $x = 0$. $(0, 0)$ only x and y intercept.

Vertical asymptotes: $x = 1$, $x = 3$.

Horizontal asymptote: The degrees of the top and bottom are equal. $y = 2/1$.

Sketch: Rightmost $x = 3$. Substitute $x = 3.1$, everything positive. To the right of $x = 3$, the curve goes to plus infinity. $(x - 3)^2$ has an even power. To the left, the curve also goes to plus infinity. The graph comes down and then back up at $x = 1$ since there are no . intercepts between $x = 3$ and $x = 1$. The power $(x - 1)$ is odd. Since one end is at plus infinity, on the left side, the curve must go to minus infinity. Then we

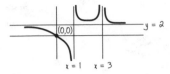

go up to (0, 0). Since the power of x^3 is odd, there is a crossing. Both ends go to the horizontal asymptote, $y = 2$. The sketch is here.

EXAMPLE 18—

$$y = \frac{x^2 - 4x + 4}{x - 1} = \frac{(x - 2)^2}{x - 1} = x - 3 + 1/(x - 1)$$

When the degree of the top is 1 more than the bottom, we need three forms: the original, the factored form (if possible), and the long-divided form.

x intercept (from factored form): (2, 0).

y intercept: $x = 0$ in original (0, −4).

Vertical asymptote: $x = 1$. No horizontal asymptote.

Oblique asymptote: $y = x − 3$ (graph this in any way you know how).

Sketch: Use factored form: $(x − 2)^4$. Substitute $x = 2.1$ into everything. y is positive. Power is even. Both ends are above intercept (2, 0); no cross. Head for $x = 1$. Right end must go to plus infinity. Since the power of $x − 1$ is odd, the other end goes to minus infinity. It then heads up to the point (0, −4). The very ends of the graph go to the line $y = x − 3$. The sketch is as shown here:

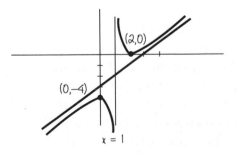

EXAMPLE 19—

$$y = \frac{x^8}{(x+3)^5} = \frac{x^8}{x^5 + \cdots} = \text{approximately } x^3$$

Intercept: (0, 0).

Vertical asymptote: $x = -3$.

Since the degree of the top is more than 1 more than the bottom, there are no horizontal or oblique asymptotes. However, we would like to know what happens to the ends. At the ends, the curve is approximately x^3. Substitute a big number, $x = 100$. $(100)^3$ is positive. The right end goes to plus infinity. Substitute a small number, $x = -100$. $(-100)^3$ is negative. The left end goes to minus infinity.

Let's do the middle (this really should be done first). (0, 0) is the rightmost. Letting $x = 0.1$, we again get y positive. As we now know, the right end goes to plus infinity. Since the power of x^8 is even, there is no crossing. The left end goes positive up to plus infinity at $x = -3$. Since the power of $(x+3)^5$ is odd and the right side goes to plus infinity, the left side goes to minus infinity. The curve gets bigger for a while, and then it turns around and heads for minus infinity again. Its sketch is shown here:

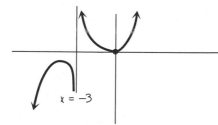

If you go through this, you will understand curve sketching and do it faster than your teacher.

This is the end of the second section on functions. There is another large, important one, but first . . .

CHAPTER 12

MODERN LOGARITHMS

We will use a modern approach to logs. ("Modern" to a mathematician means not more than 50 years behind the times.) We will not do calculations with logs (calculations involving characteristics and mantissas). This is no longer needed because we have calculators. What is needed is a thorough understanding of the laws of logarithms and certain problems that can be solved only with logs.

DEFINITION

$\log_b x = y$ (log of x to the base b is y) if $b^y = x$.

NOTE

y, the answer, which is also the log itself, is an exponent ($b^y = x$). The laws of logarithms are the laws of exponents, as we will see.

EXAMPLE 1—

Show that $\log_5 25 = 2$ is true.

Using the definition, $\log_5 25 = 2$ means $5^2 = 25$.

EXAMPLE 2—

Write $4^3 = 64$ in log form.

NOTE

This definition must be practiced!!!!!

The base is 4, the exponent is 3, the answer is 64. So we get $\log_4 64 = 3$.

EXAMPLE 3—

Find $\log_{16} 64$.

Let $\log_{16} 64 = x$. Use the definition of log since this is all we know at this point.

$$16^x = 64 \qquad (2^4)^x = 2^6 \qquad 2^{4x} = 2^6$$

Since the bases are equal, the exponents are equal. $4x = 6$. $x = 3/2$.

EXAMPLE 4—

Solve for x: $\log_9 x = -3/2$.

$9^{-3/2} = x$. (Review fractional exponents.)

$$x = 9^{-3/2} = \frac{1}{9^{3/2}} = 1/(9^{1/2})^3 = \frac{1}{3^3} = \frac{1}{27}$$

EXAMPLE 5—

Solve for x: $\log_x 32 = 5$.

$x^5 = 32$. So $x = 32^{1/5}$. $x = 2$.

We will now discuss what the base can be and the domain and range of logs.

Can the base be negative? No, since $(-2)^{1/2}$ is imaginary. Can the base be 0? No, since 0^n is either 0 or undefined (if n is 0 or negative). Are there any other exclusions? Yes. b cannot be 1 since 1^n is always 1.

The base can be any positive number other than 1. The base could be $2^{1/2}$, but it won't do you much good since there are no tables or calculators for this. The two most common bases are 10, because we have 10 fingers, and e, a number that comes up a lot in later math.

1. e is approximately 2.7. More exactly? On your calculator press 1, inv, 1n.

2. $\log_{10}$ = log.

3. $\log_e$ = ln (ln is called the *natural logarithm*).

DEFINITIONS

Range of logs (the y values): A log, y, is an exponent. Exponents can be positive, 0, and negative. The range is all real numbers.

Domain of logs (the x values): Since the base is positive, whether the exponent is positive, 0, or negative, the answer x will always be positive. So the domain is all numbers bigger than 0. (Remember, *negative exponent* means reciprocal.)

LAW I

$\log_b xy = \log_b x + \log_b y$

EXAMPLE 6—

$\log 6 = \log (2)(3) = \log 2 + \log 3$

LAW 2

$\log_b (m/n) = \log_b m - \log_b n$

EXAMPLE 7—

$\log (4/3) = \log 4 - \log 3$

LAW 3

$\log_b x^p = p \log_b x$

EXAMPLE 8—

$\log 32 = \log 2^5 = 5 \log 2$

These three laws are the most important. In calculus, if you can do the next problem, you know 50% of what you need.

EXAMPLE 9—

Simplify with no exponents:

$$\ln \left(\frac{a^4 \sqrt{b}}{c^6 d^7} \right)$$

Add all the logs on the top, subtract all the logs on the bottom, and the exponents come down and multiply the specific log.

SOLUTION—

$4 \ln a + \frac{1}{2} \ln b - 6 \ln c - 7 \ln d$

EXAMPLE 10—

Write as a single log:

$5 \log h - 7 \log c - 8 \log p - 4 \log j + (3/4) \log k + \log v$

Everything with a plus sign is multiplied in the numerator; everything with a minus sign is multiplied in the denominator; coefficients come up as exponents.

SOLUTION—

$$\log \left(\frac{h^5 k^{3/4} v}{c^7 p^8 j^4} \right)$$

LAW 4

$\log_b b = 1$ since $b^1 = b$. $\log_7 7 = \log 10 = \ln e = 1$.

LAW 5

$\log_b 1 = 0$ since $b^0 = 1$. $\log_9 1 = \log 1 = \ln 1 = 0$.

LAW 6
The log is an increasing function. If m < n, then log m < log n. We know log 2 < log 3 since 2 < 3.

LAW 7
The log is 1:1. That is, if $\log_b x = \log_b y$, then x = y.

EXAMPLE 11—

$\log_5 (x^{1/2}) = \log_5 (2x - 3)$.

By 1:1,

$x^{1/2} = 2x - 3$

> We must square both sides. In doing this, we might introduce extraneous solutions. We must check them back in the original equation.

$(x^{1/2})^2 = (2x - 3)^2 \quad$ or $\quad x = 4x^2 - 12x + 9$

$4x^2 - 13x + 9 = 0$

> Now factor.

$(4x - 9)(x - 1) = 0 \quad x = 9/4 \quad$ or $\quad x = 1$

Check x = 9/4:

$\log_5 (9/4)^{1/2} \overset{?}{=} \log_5 [2(9/4) - 3]$

$\log_5 (3/2) = \log_5 (3/2)$. Yes, this one checks.

Check x = 1:

$\log_5 (1)^{1/2} \overset{?}{=} \log_5 [2(1) - 3]$

$\log_5 1 \ne \log_5 (-1)$ since there is no log of a negative number.

The only solution is x = 9/4.

LAW 8
$b^{\log_b x}$ is a weird way of writing x. $e^{\ln x} = x$.

LAW 9
$\log_b b^x = x$. $\ln e^x = x$.

LAW 10

$$\log_d c = \frac{\log_b c}{\log_b d}$$

$$\log_5 7 = \frac{\log_{10} 7}{\log_{10} 5}$$

This is long division, which of course we don't have to do now since we have calculators.

The following are problems involving logs that you should be able to do.

EXAMPLE 12

$$(4) \cdot 3^{x+2} = 28$$

Isolate the exponent. Divide both sides by 4.

$$3^{x+2} = 7$$

Now take logs to get x "off the exponent."

At this point, this is an elementary algebra problem. You must solve for x. Remember, log 3 and log 7 are numbers.

$$(x + 2) \log 3 = \log 7$$

$$x \log 3 + 2 \log 3 = \log 7$$

Using a calculator, we get x = −0.23.

$$x = \frac{\log 7 - 2 \log 3}{\log 3}$$

EXAMPLE 13

$$5^{3x-6} = 7^{8x+9}$$

Take logs.

$$(3x - 6) \log 5 = (8x + 9) \log 7$$

$$3x \log 5 - 6 \log 5 = 8x \log 7 + 9 \log 7$$

$$3x \log 5 - 8x \log 7 = 6 \log 5 + 9 \log 7$$

x(3 log 5 − 8 log 7) = 6 log 5 + 9 log 7

So $x = \dfrac{6 \log 5 + 9 \log 7}{3 \log 5 - 8 \log 7}$

EXAMPLE 14—

$\log_2 x + \log_2 (x - 2) = 3$

log m + log n = log mn. Then define logs:

$\log_2 x(x - 2) = 3$ or $2^3 = x(x - 2)$ or $x^2 - 2x - 8 = 0$

or $(x - 4)(x + 2) = 0$

x = −2 is rejected since you cannot take the log of a negative.

x = 4 is OK since $\log_2 4 + \log_2 (4 - 2) = \log_2 4 + \log_2 2 = 2 + 1 = 3$.

EXAMPLE 15—

log (x − 3) − log 4 = 2

log m − log n = log m/n. Then define logs, remembering that *log* means the base is 10:

$\log_{10} (x - 3)/4 = 2$ or $(x - 3)/4 = 10^2$. x − 3 = 400. So x = 403. Not too bad, is it?

Sometimes you are given problems involving compound interest. The formula is given by

$A = P(1 + r/n)^{nt}$

where A = amount at the end; P = principal, amount you put in; r = interest rate per year; n = number of times compounded a year; and t = years.

EXAMPLE 16—

Say we invest $1000 at 20% interest. If it is compounded four times a year, when will we have $4000?

$A = 4000$, $P = 1000$, $r = 0.20$, $n = 4$, $t = ????????$

SOLUTION—

Substitute $4000 = 1000(1 + 0.20/4)^{4t}$. $4000 = 1000(1.05)^{4t}$. Divide by 1000 and take logs:

$4 = (1.05)^{4t}$ $\log 4 = 4t \log 1.05$ $t = \dfrac{\log 4}{4 \log 1.05}$

$t = 7.1$ years

This means in 7.1 years our investment is multiplied by 4. Now if we can only find a 20% safe return!

EXAMPLE 17—

If the half-life of a radioactive substance is 8 days and we start with 10 pounds of undecayed substance, when will there be 3 pounds of active material left?

In calculus you will derive this. At that time, it will not be hard. For now, however, the formula is $A = A_0(½)^{t/8}$, where A = amount now, A_0 = amount in the beginning ($t = 0$), t = time, and the denominator of t is the time of the half-life.

Let us check the formula: $t = 8$, $A = 10(½)^{8/8} = 10(½) = 5$, one-half as much in 8 days!

Let us do the problem to find out how many days for $A = 3$ pounds left:

Divide by 10, take logs, and solve for t.

$3 = 10(½)^{t/8}$

$0.3 = (½)^{t/8}$

$\log 0.3 = (t/8) \log 0.5$

$t = (8 \log 0.3)/\log 0.5 = 13.9$ days   (by calculator)

If you know these problems, you will be all set not only for precalc but for calc.

CHAPTER 13

PARABOLAS II, ELLIPSES, AND HYPERBOLAS

We will now discuss these three curves the way they are discussed in calc books and some precalc books.

The important way to study these curves is to relate the equation to the picture. If you do this, this entire chapter will become much easier.

DEFINITION

Parabola: The set of all points equidistant from a point, called a *focus,* and a line, called the *directrix.* F is the focus. The point V, the *vertex,* is the closest point to the directrix.

NOTE

According to the definition $\overline{FV} = \overline{VR}$, $\overline{FP_1} = \overline{P_1R_1}$, $\overline{FP_2} = \overline{P_2R_2}$, etc.

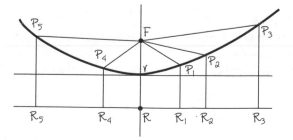

Let's do this development algebraically.

Let the vertex V be at (0, 0) and the focus F be at (0, c). The directrix is at y = −c. Let P(x, y) be any point on the parabola. The definition of the parabola says $\overline{FP} = \overline{PQ}$. Everything on $\overline{PQ}$ has the same x value, and everything on y = −c has the same y value. Point Q has to be (x, −c). On $\overline{PQ}$, since the x values are the same, the length of the line is y − (−c). Using the distance formula to get $\overline{PF}$ and setting the two segments squared equal to each other, we get . . . $(x − 0)^2 + (y − c)^2 = (y + c)^2$. Squaring, we get $x^2 + y^2 − 2cy + c^2 = y^2 + 2cy + c^2$. Simplifying, we get $x^2 = 4cy$.

Here is a chart that will be helpful in relating the equation to the picture.

Vertex	Focus	Directrix	Equation	Picture	Comment
(0, 0)	(0, c)	y = −c	$x^2 = 4cy$		The original derivation
(0, 0)	(0, −c)	y = c	$x^2 = −4cy$		y replaced by −y
(0, 0)	(c, 0)	x = −c	$y^2 = 4cx$		x, y interchange in top line
(0, 0)	(−c, 0)	x = c	$y^2 = −4cx$		x replaced by −x in the above line

EXAMPLE 1—

$y^2 = −7x$. Sketch; label vertex, focus, directrix.

The chart tells us the picture is the last line. Now let 4c equal 7, ignoring the minus sign. c = 7/4. Vertex is (0, 0). Focus is (−7/4, 0) because it is on the negative x axis. The directrix is x = +7/4, positive because it is to the right and x = 7/4 since it is a vertical line. Easy, isn't it?

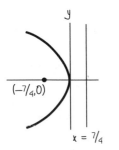

EXAMPLE 2—

Sketch $(y − 3)^2 = −7(x + 2)$.

To understand this curve, we look at the difference between $x^2 + y^2 = 25$ and $(x - 3)^2 + (y + 6)^2 = 25$. Has the shape changed? No. Has the radius changed? No. The only thing that has changed is its position. The center is now at the point (3, –6) instead of at the point (0, 0).

In the case of our little parabola, it is the vertex that has changed. V = (–2, 3). (Remember x is always first.) 4c is still equal to 7. c = 7/4, but F is (–2 – 7/4, 3), 7/4 to the left of the vertex. The directrix is x = –2 + 7/4. I do not do the arithmetic so that you know where the numbers come from.

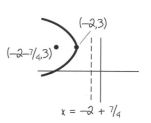

$$x = -2 + 7/4$$

EXAMPLE 3—

Sketch the parabola $2x^2 + 8x + 6y + 10 = 0$.

$2x^2 + 8x + 6y + 10 = 0$

Divide by coefficient of x^2 or y^2.

$x^2 + 4x + 3y + 5 = 0$

On the left get all the terms related to the square term; everything else on the other side.

$x^2 + 4x = -3y - 5$

Complete the square; add to both sides.

$x^2 + 4x + 4 = -3y - 5 + 4$

Factor; do the arithmetic.

$(x + 2)^2 = -3y - 1$

Weird last step. No matter what the coefficient on the right, factor it all out even if it results in a fraction in the parentheses.

$$(x + 2)^2 = -3\left(y + \frac{1}{3}\right)$$

Sketch; the picture is below. V(–2, –1/3). 4c = 3. c = 3/4. F(–2, –1/3 – 3/4). Directrix y = –1/3 + 3/4.

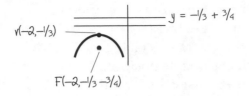

DEFINITION

Ellipse: Algebraically, the ellipse is defined as $\overline{PF_1} + \overline{PF_2} = 2a$. $2a > 2c$. $2c$ is the distance between the *foci*. a will be determined later. P is any point on the ellipse.

To paraphrase the definition, given two points called *foci*, we draw a line from one of the points to any point on the curve and then from that point on the curve to the other focus; if the sum of the two lengths always adds to the same number, $2a$, the figure formed will be an ellipse.

I know you'd all desperately like to draw an ellipse. Take a nonelastic string. Attach thumbtacks at either end. Take a pencil and stretch the string as far as it goes. Go 360°. You will have traced an ellipse.

Some of you have seen the equation of an ellipse, but few of you have seen its derivation. It is an excellent algebraic exercise for you to try. You will see that there is a lot of algebra behind a very simple equation.

Use the distance formula.	$\overline{PF_1} + \overline{PF_2} = 2a$
Isolate one square root, and square both sides.	$\sqrt{(x-(-c))^2 + (y-0)^2} + \sqrt{(x-c)^2 + (y-0)^2} = 2a$
Do the algebra.	$[\sqrt{(x+c)^2 + y^2}]^2 = [2a - \sqrt{(x-c)^2 + y^2}]^2$
Combine like terms; isolate the radical.	$x^2 + 2cx + c^2 + y^2 = 4a^2 + x^2 - 2cx + c^2 + y^2$ $- 4a\sqrt{(x-c)^2 + y^2}$

$$4a\sqrt{(x-c)^2 + y^2} = 4a^2 - 4cx$$

Divide by 4; again square both sides.

$$[a\sqrt{(x-c)^2 + y^2}]^2 = (4a^2 - 4cx)^2$$

Do the algebra.

$$a^2(x^2 - 2cx + c^2 + y^2) = a^4 - 2a^2cx + c^2x^2$$

$$\text{or} \quad a^2x^2 - c^2x^2 + a^2y^2 = a^4 - a^2c^2$$

$$\frac{(a^2 - c^2)x^2}{(a^2 - c^2)a^2} + \frac{a^2y^2}{a^2(a^2 - c^2)} = \frac{a^2(a^2 - c^2)}{a^2(a^2 - c^2)}$$

Factor out x^2 on the left two terms and a^2 from the right two terms; divide by $(a^2 - c^2) \cdot a^2$.

We get

$$\frac{x^2}{a^2} + \frac{y^2}{a^2 - c^2} = 1 \quad \text{or} \quad \frac{x^2}{a^2} + \frac{y^2}{b^2} = 1$$

letting $a^2 - c^2 = b^2$. Whew!!!!!!

We are still not finished. Let's find out what a is and what b is. $\overline{F_1P} + \overline{PF_2} = 2a$, where P is any point on the ellipse. Let T be the point P. $\overline{F_1T} + \overline{TF_2} = 2a$. By symmetry, $\overline{F_1T} = \overline{TF_2}$. So $\overline{F_1T} = \overline{TF_2} = a$. Since $a^2 - c^2 = b^2$, $\overline{OT} = \overline{OT'} = b$. The coordinates of T are $(0, b)$; T' is $(0, -b)$.

We would like to find the coordinates of U (and U'), but the letters a, b, and c are used up. Oh well, let's see what happens. $\overline{F_2U} + \overline{UF_1} = 2a$. $\overline{F_2U} = x - c$. $\overline{UF_1} = x + c$. $x - c + x + c = 2a$. So $2x = 2a$; $x = a$. U is $(a, 0)$; U' is $(-a, 0)$.

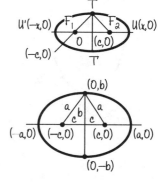

c = half the distance between the foci

b = length of the semiminor axis (*semi* means half; *minor* means smaller; *axis* means line)

a = length of semimajor axis = distance from a focus to a minor vertex

$(\pm a, 0)$—major vertices

$(0, \pm b)$—minor vertices

$(\pm c, 0)$—foci, always located on the major axis

Although the derivation was long, sketching should be short.

EXAMPLE 4—

Sketch $x^2/7 + y^2/5 = 1$.

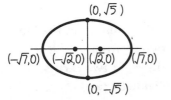

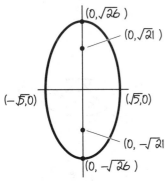

In the case of the ellipse, the longer axis is indicated by the larger number under x^2 or y^2. Try not to remember a or b. Remember the picture! This is longer in the x direction. $y = 0$. Major vertices: $(\pm 7^{1/2}, 0)$ $x = 0$. Minor vertices: $(0, \pm 5^{1/2})$. $c = (7 - 5)^{1/2}$. Foci: $(\pm 2^{1/2}, 0)$.

EXAMPLE 5—

$$\frac{x^2}{5} + \frac{y^2}{26} = 1$$

Longer in the y direction. Major vertices: $(0, \pm 26^{1/2})$. Minor vertices: $(0, \pm 5^{1/2})$. $c = (26 - 5)^{1/2}$. Foci: $(0, \pm 21^{1/2})$. Foci are always on the longer axis.

EXAMPLE 6—

$$\frac{(x - 6)^2}{7} + \frac{(y + 4)^2}{5} = 1$$

This is the same basic example as Example 4, except the middle is no longer at $(0, 0)$. It is at the point $(6, -4)$. Major vertices: $(6 \pm 7^{1/2}, -4)$. Minor vertices: $(6, -4 \pm 5^{1/2})$. Foci: $(6 \pm 2^{1/2}, -4)$.

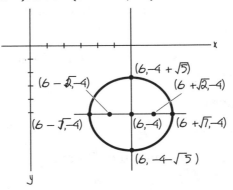

NOTE

The numbers found in Example 4 are added and subtracted from the appropriate coordinate of the center (6, −4). Also note that weird numbers were purposely chosen so that you could see where the numbers came from.

EXAMPLE 6 REVISITED—

$$\frac{(x-6)^2}{7} + \frac{(y+4)^2}{5} = 1$$

The center is again at the point (6, −4). The vertices are directly east-west or north-south of (6, −4). All points east-west of (6, −4) must have the same y value, y = −4. Soooooo

$$\frac{(x-6)^2}{7} + \frac{(-4+4)^2}{5} = 1$$

$$\frac{(x-6)^2}{7} + 0 = 1$$

$$(x-6)^2 = 7$$

$$x - 6 = \pm 7^{1/2}$$

Soooooo x = 6 ± 7^{1/2} and the Major vertices are (6 ± 7^{1/2}, −4). Points north-south of (6, −4) have the same x values, x = 6.

$$\frac{(6-6)^2}{7} + \frac{(y+4)^2}{5} = 1$$

$$0 + \frac{(y+4)^2}{5} = 1$$

$$(y+4)^2 = 5$$

$$y + 4 = \pm 5^{1/2}$$

Therefore, $y = -4 \pm 5^{1/2}$ and the Minor vertices are $(6, -4 \pm 5^{1/2})$. For the same reasons, the Foci, always on the major axis, are $(6 \pm 2^{1/2}, -4)$. The sketch is, of course, the same.

EXAMPLE 7—

Sketch and discuss $4x^2 + 5y^2 + 30y - 40x + 45 = 0$.

Like the parabola and circle, we must complete the square, only a little differently.

Group the x and y terms; number to the other side.

$$4x^2 + 5y^2 + 30y - 40x + 45 = 0$$

Factor out coefficients of x^2 and y^2; complete the square in the parentheses; add the number term inside the parentheses multiplied by the number outside the parentheses to each side—with both x and y. Then do the arithmetic and divide by 100 to get 1 on the right.

$$4x^2 - 40x + 5y^2 + 30y = -45$$

$$4\left[x^2 - 10x + \left(\frac{-10}{2}\right)^2\right] + 5\left[y^2 + 6y + \left(\frac{6}{2}\right)^2\right]$$

$$= -45 + 4\left(\frac{-10}{2}\right)^2 + 5\left(\frac{6}{2}\right)^2$$

$$\frac{4(x-5)^2}{100} + \frac{5(y+3)^2}{100} = \frac{100}{100}$$

or $\dfrac{(x-5)^2}{25} + \dfrac{(y+3)^2}{20} = 1$

The center is (5, −3). Under the $(x-5)^2$ term is larger and $25^{1/2}$ to the left and right of the center. Major vertices are $(5 \pm 25^{1/2}, -3)$. Under the $(y+3)^2$, $20^{1/2}$ above and below the center. Minor vertices are $(5, -3 \pm 20^{1/2})$. $c = (25-20)^{1/2}$. Foci, on the larger axis, are $(5 \pm 5^{1/2}, -3)$.

Of course you should use 5 instead of $25^{1/2}$, but I left $25^{1/2}$ to show you where the 5 came from.

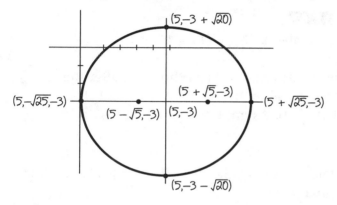

DEFINITION

Hyperbola: The set of all points P such that $\overline{F_1P} - \overline{PF_2} = 2a$.

The derivation is nearly the same as for the ellipse. Once is enough!!!!! The equation is $x^2/a^2 - y^2/b^2 = 1$, where $a^2 + b^2 = c^2$. $(\pm a, 0)$—Transverse vertices. $(\pm c, 0)$—foci. Asymptotes: $y = \pm(b/a)x$. Slopes of the lines are the square root of the number under the y^2 term over the square root of the number under the x^2 term. The shape of the curve depends on the location of the minus sign, *not* the largeness of the numbers under the x^2 or y^2 term.

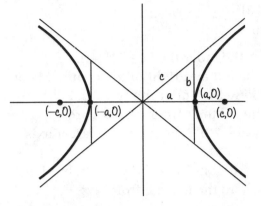

EXAMPLE 8—

Sketch and label $x^2/7 - y^2/11 = 1$.

Transverse vertices: $y = 0$ (set the letter after the minus sign equal to 0). $(\pm 7^{1/2}, 0)$. $c = (7 + 11)^{1/2}$. Foci: $(\pm 18^{1/2}, 0)$. Asymptotes: $y = \pm(11^{1/2}/7^{1/2})x$.

NOTE

Curve does not hit y axis. If $x = 0$, $y = \pm(-11)^{1/2}$, which are imaginary.

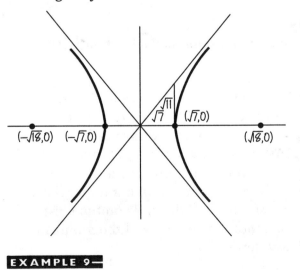

EXAMPLE 9—

Sketch and discuss $y^2/5 - x^2/9 = 1$.

Let x = 0. Transverse vertices: $(0, \pm 5^{1/2})$. $c = (5 + 9)^{1/2}$. Foci: $(0, \pm 14^{1/2})$. Asymptotes: $y = \pm(5^{1/2}/9^{1/2})x$. The sketch is as shown here.

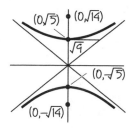

EXAMPLE 10—

Sketch and discuss $(y - 6)^2/5 - (x + 7)^2/9 = 1$.

This is the same as Example 9, except the "center" of the hyperbola, where the asymptotes cross, is no longer at (0, 0). The center is (−7, 6). $(x + 7)^2 = 0$; then $a = 5^{1/2}$, $5^{1/2}$ above and below the center. $c = (5 + 9)^{1/2}$. Foci are $14^{1/2}$ above and below the center. $V(-7, 6 \pm 5^{1/2})$. $F(-7, 6 \pm 14^{1/2})$. Asymptotes: $y - 6 = \pm(5^{1/2}/9^{1/2})(x + 7)$.

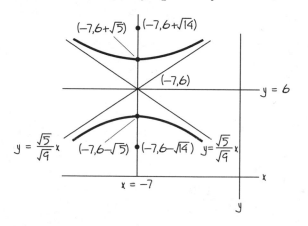

EXAMPLE 10 REVISITED—

$$\frac{(y - 6)^2}{5} - \frac{(x + 7)^2}{9} = 1$$

Just as with the ellipse, let's clarify the location of the vertices. In this case, the vertices occur north-south of the center (−7, 6). So the x value of the vertices is the same as the center, (−7, 6). x = −7. Substituting, we get

$$\frac{(y - 6)^2}{5} - \frac{(-7 + 7)^2}{9} = 1$$

$(y - 6)^2 = 5$

$y = 6 \pm 5^{1/2}$ and the Vertices are $(-7, 6 \pm 5^{1/2})$. For the same reason, the Foci are $(-7, 6 \pm 14^{1/2})$. The asymptotes and the sketch are the same!

EXAMPLE 11—

Sketch and discuss $25x^2 - 4y^2 + 50x - 12y + 116 = 0$.

For the last time, we will complete the square, again a little differently than the other times. We will use exactly the same steps as for the ellipse, except for the minus sign.

$$25x^2 - 4y^2 + 50x - 12y + 116 = 0$$

$$25x^2 + 50x - 4y^2 - 12y = -116$$

$$25\left[x^2 + 2x + \left(\frac{2}{2}\right)^2\right] - 4\left[y^2 + 3y + \left(\frac{3}{2}\right)^2\right]$$

$$= -116 + 25\left(\frac{2}{2}\right)^2 - 4\left(\frac{3}{2}\right)^2$$

$$\frac{25(x + 1)^2}{-100} - \frac{4\left(y + \frac{3}{2}\right)^2}{-100} = \frac{-100}{-100}$$

$$\frac{\left(y + \frac{3}{2}\right)^2}{25} - \frac{(x + 1)^2}{4} = 1$$

The center is $(-1, -3/2)$. $V(-1, -3/2 \pm 25^{1/2})$. $F(-1, -3/2 \pm 29^{1/2})$. Asymptotes: $y + 3/2 = \pm(25^{1/2}/4^{1/2})(x + 1)$.

Sometimes we have a puzzle. Given some information, can we find the equation? You must always draw the picture and relate the picture to its equation.

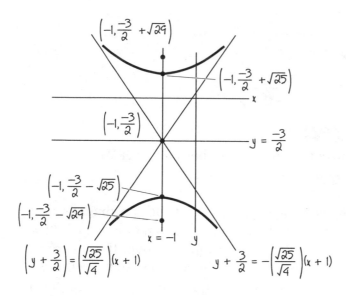

$$\left(y + \frac{3}{2}\right) = \left(\frac{\sqrt{25}}{\sqrt{4}}\right)(x + 1)$$

$$y + \frac{3}{2} = -\left(\frac{\sqrt{25}}{\sqrt{4}}\right)(x + 1)$$

EXAMPLE 12—

Find the equation of the parabola with Focus (1, 3), directrix x = 11.

Drawing F and the directrix, the picture must be the one shown here. The vertex is halfway between the x numbers. So x = (11 + 1)/2 = 6. V(6, 3). c = the distance between V and F = 5. The equation is $(y - 3)^2 = -4c(x - 6)$ = $-20(x - 6)$. Remember, the minus sign is from the shape and c is always positive for these problems.

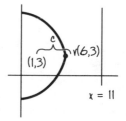

EXAMPLE 13—

Vertices are (2, 3) and (12, 3), and there is one Focus (11, 3). Find the equation of the ellipse.

The two vertices give the center [(12 + 2)/2, 3] = (7, 3). F(11, 3). $(x - 7)^2/a^2 + (y - 3)^2/b^2 = 1$. a = 12 − 7 = 5. c = 11 − 7 = 4. $a^2 - b^2 = c^2$. $5^2 - b^2 = 4^2$. $b^2 = 9$ (no need for b). $(x - 7)^2/25 + (y - 3)^2/9 = 1$.

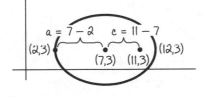

EXAMPLE 14—

Find the equation of the hyperbola with Vertices
(0, ±6) and Asymptotes y = ±(3/2)x.

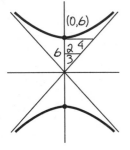

V(0, ±6) says the center is (0, 0) and the shape is
$y^2/36 - x^2/a^2 = 1$. The slope of the asymptotes is 3/2 =
square root of the number under y^2 over the square
root of the number under the x^2 term. So 3/2 = 6/a. So
a = 4. So $a^2 = 4^2 = 16$. The equation is $y^2/36 - x^2/16 = 1$.

This kind of question is shorter in length, but it does
take practice. So practice!!!

SYSTEMS OF NONLINEAR EQUATIONS

We are now going to solve systems of nonlinear (not
lines) equations. Some of the curves may be conics,
some may be conics in disguised forms, and some are
not conics at all. We will use the elimination method
and the substitution method to solve linear equations
in two or three unknowns. Sometimes we will use
both methods and sometimes there are tricks you
wouldn't believe. But that is why I'm putting this
section here.

EXAMPLE 15—

Solve for all x and y in common.

$y^2 - x^2 = 27$

$y - x = 3$

We will solve this by substitution. We solve for x or y in the second equation (y is the slightly better choice here) and substitute it in the first equation.

$y = x + 3$

So $(x + 3)^2 - x^2 = 27$

$$6x + 9 = 27$$

$$6x = 18$$

$$x = 3$$

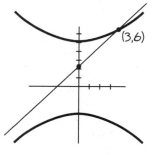

Substituting in the second equation, we get $y = 3 + 3 = 6$. The point is $(3, 6)$. The picture looks like the one shown on this page.

NOTE 1
If one equation is a line and the other is a circle, ellipse, parabola, or hyperbola, there may be at most two and as few as no answers (two or one or zero).

NOTE 2
Always substitute the values into the easier equation: first, it's easier; second, there is less chance of an extraneous solution (nonanswer).

NOTE 3
There is a neat way of doing this equation pair:

$y - x = 3$ and $y^2 - x^2 = (y - x)(y + x) = 27$

$3(y + x) = 27$

$$y + x = 9$$

$$y - x = 3$$

Adding, $2y = 12$.

$y = 6$

Substituting, we get $x = 3$. Nice!

NOTE

It is a trick that sometimes occurs in the SAT 1.

EXAMPLE 16—

Solve for all x and y.

$x^2 + y^2 = 16$

$x^2 = 15y$

We will use the substitution method, but instead of substituting for y, we will substitute for x^2.

$$15y + y^2 = 16$$

$$y^2 + 15y - 16 = 0$$

$$(y + 16)(y - 1) = 0$$

$$y = -16$$

$$y = 1$$

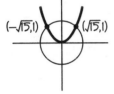

If $y = -16$, then $x^2 = 15(-16)$, which has no real solution since x^2 is negative. If $y = 1$, then $x^2 = 15$ and $x = \pm(15)^{1/2}$. The solutions are $(15^{1/2}, 1)$ and $(-15^{1/2}, 1)$.

NOTE

This is the intersection of a parabola and a circle. In this case there are two answers. There could be as many as four and as few as zero.

ALSO NOTE

It is impossible to find the solution exactly using a graph since $15^{1/2}$ is irrational.

EXAMPLE 17—

Solve for all x and y. We will use the elimination method.

$$4x^2 + 5y^2 = 22$$

$$7x^2 - 2y^2 = 17$$

$$2(4x^2 + 5y^2) = 22(2)$$

$$5(7x^2 - 2y^2) = 17(5)$$

$$43x^2 = 129$$

$$x^2 = 3$$

$$x = \pm3^{1/2}$$

$$4(\pm3^{1/2})^2 + 5y^2 = 22$$

$$12 + 5y^2 = 22$$

$$y^2 = 2$$

$$y = \pm2^{1/2}$$

For each of the two x values, there are two y values, four solutions in all: $(3^{1/2}, 2^{1/2})$, $(3^{1/2}, -2^{1/2})$, $(-3^{1/2}, 2^{1/2})$, $(-3^{1/2}, -2^{1/2})$. The picture looks like this (it is not drawn to scale; too messy!!)

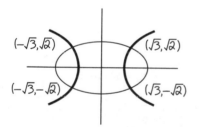

Now let's try some different examples. No more pictures.

EXAMPLE 18—

Solve for all x and y in common.

$$\frac{4}{x} + \frac{5}{y} = 2$$

$$\frac{3}{x} - \frac{7}{y} = 23$$

We can't clear fractions since on the right side we would get xy. But we can let $u = 1/x$ and $v = 1/y$. The equations becommmme . . .

Now eliminate by multiplying the top by 7 and the bottom by 5.

$$4u + 5v = 2$$

$$3u - 7v = 23$$

$$28u + 35v = 14$$

$$15u - 35v = 115$$

$$43u = 129$$

Substituting, v = –2.

$$u = 3$$

But we are not finished. $x = 1/u = 1/3$ and $y = 1/v = -\frac{1}{2}$.

EXAMPLE 19—

Solve for all x and y in common.

$$y = 3^x$$

$$y = 3^{2x} - 72$$

First we substitute since y = y.

$$3^{2x} - 72 = 3^x$$

Let $u = 3^x$, so $u^2 = 3^{2x}$. Neat, huh?!

$$3^{2x} - 3^x - 72 = 0$$

$$u^2 - u - 72 = 0$$

$$(u - 9)(u + 8) = 0$$

$$u = 9 \text{ or } u = -8$$

But $u = 3^x$. $3^x = 9$. So $x = 2$. $3^x = -8$. No solution since 3^x must be positive.

NOTE

If $3^x = 10$, for example, we would have to take logs or ln's to get the answer.

EXAMPLE 20—

Solve for all x and y in common.

$x^2 + xy + y^2 = 37$

$x^2 + y^2 = 25$

We will eliminate x^2 and y^2 by subtracting. Solve for y and substitute in the second equation. A really long problem. Subtracting, we get $xy = 12$:

$$y = \frac{12}{x}$$

So $x^2 + (12/x)^2 = 25$

$$x^2 + \frac{144}{x^2} = 25 \qquad\qquad \textbf{Multiply by } x^2.$$

$x^4 - 25x^2 + 144 = 0$

$(x^2 - 9)(x^2 - 16) = (x - 3)(x + 3)(x - 4)(x + 4) = 0$

So x = 3, –3, 4, and –4. We must substitute into $x^2 + y^2 = 25$.

$(3)^2 + y^2 = 25$. $y^2 = 16$. $y = \pm 4$. But y = –4 does not check in the other equation. So (3, 4) is one solution. A similar thing happens with each of the other x values. The answers are (3, 4), (4, 3), (–3, –4), and (–4, –3).

The next examples are weirder.

EXAMPLE 21—

Solve for x, y, and z, in terms of a, b, and c, where a, b, and c are positive numbers.

A. $yz = a^2$

B. $xz = b^2$

C. $xy = c^2$

We will divide A by B and substitute in C.

$$\frac{yz}{xz} = \frac{y}{x} = \frac{a^2}{b^2}$$

So $y = a^2 x/b^2$.

C becomes $x(a^2 x/b^2) = c^2$. $x^2 = b^2 c^2/a^2$. $x = bc/a$. Similarly, $y = ac/b$ and $z = ab/c$.

EXAMPLE 22—

Solve for x, y, and z, in terms of a, b, and c, where a, b, and c are numbers.

$$x(x + y + z) = a^2$$

$$y(x + y + z) = b^2$$

$$z(x + y + z) = c^2$$

This is really easy . . . iffff you see the trick. Just add them up.

$$x(x + y + z) + y(x + y + z) + z(x + y + z)$$
$$= (x + y + z)(x + y + z) = (x + y + z)^2 = a^2 + b^2 + c^2$$

So $(x + y + z) = [a^2 + b^2 + c^2]^{1/2}$. $x = a^2/(x + y + z) = a^2/[a^2 + b^2 + c^2]^{1/2}$. Similarly, $y = b^2/[a^2 + b^2 + c^2]^{1/2}$ and $z = c^2/[a^2 + b^2 + c^2]^{1/2}$.

And now, perhaps the weirdest and hardest of all!!!

EXAMPLE 23—

Solve for x and y.

$$\ln\left(\frac{x}{y}\right) = \frac{\ln x}{\ln y} \qquad x^6 = y^8$$

THE SOLUTION

A really tough one. You must know your logs. We multiply by 6.

$$6 \ln\left(\frac{x}{y}\right) = 6\,\frac{\ln x}{\ln y}$$

$$\ln\left(\frac{x}{y}\right)^6 = \frac{\ln x^6}{\ln y}$$

$$\ln\left(\frac{x^6}{y^6}\right) = \frac{\ln x^6}{\ln y}$$

We know $x^6 = y^8$.

$$\ln\left(\frac{y^8}{y^6}\right) = \frac{\ln y^8}{\ln y} = \frac{8 \ln y}{\ln y}$$

$$\ln y^2 = 8$$

$$2 \ln y = 8$$

$$\ln y = 4 \qquad y = e^4$$

$$x^6 = y^8 = (e^4)^8 = e^{32} \qquad x = e^{32/6} = e^{16/3}$$

Wow!! Let's check. I still don't believe it!

$$\ln\left(\frac{e^{16/3}}{e^4}\right) \overset{?}{=} \frac{\ln\left(e^{16/3}\right)}{\ln\left(e^4\right)}$$

$$\ln e^{16/3 - 4} \overset{?}{=} \frac{(16/3) \ln e}{4 \ln e} \qquad \ln e = 1$$

$$\ln e^{4/3} \overset{?}{=} \frac{1}{4}\left(\frac{16}{3}\right)$$

$$4/3 \ln e = 4/3 = 4/3$$

It's true! What a struggle! I hope you don't get too many like this one!!

CHAPTER 14

WRITING FUNCTIONS OF x, THE ALGEBRAIC PART OF CALCULUS WORD PROBLEMS

In calc I, the most difficult section for the largest number of people is the section on word problems. It has very little to do with the calculus part of the problems. Ninety percent of the time the calculus is very, very easy. It is the algebraic setting up of the problem that causes all the trouble. Most textbooks thoroughly neglect this area of algebra, and even if the book does have these problems, most of the time the teacher skips this section. In calculus you are then faced with both the algebra and calculus, which is too much for most students. We will try to make word problems a lot less painful.

Here are the keys to the word problems:

1. Don't panic.

2. Don't panic!

3. Don't panic!!!! These first three rules are very important.

4. Find out what you are solving for. This seems like a silly thing to say, but there is usually more than one possibility in the problem.

5. If you can draw a picture of the problem, draw it. On the picture, label the unknowns in terms of one variable if possible, or two variables otherwise.

6. If you have trouble with these problems, do not worry about it. Equally important, do not quit. Read the solutions over and over. With hard work, you will get the knack of doing these problems.

Let's start with an easy one.

EXAMPLE 1—

If the height of a triangle is three-fourths of the base,

A. Write the area of the triangle in terms of the base.

B. Write the area of the triangle in terms of the height.

In this problem, a picture is not really necessary. This problem tells you you are looking for the area of a triangle. (You must know these formulas. They are, of course, given earlier in this book.) $A = \frac{1}{2}bh$. The problem also tells you the height is three-fourths of the base. In symbols, $h = (3/4)b$.

A. $A = \frac{1}{2}bh = \frac{1}{2}b[(3/4)b] = (3/8)b^2$.

B. For this part, we must get b in terms of h. $h = (3/4)b$. $4h = 3b$. $b = (4/3)h$. So $A = \frac{1}{2}bh = \frac{1}{2}[(4/3)h]h = (2/3)h^2$.

OK. This one isn't too bad. Let's try another not-too-bad problem.

EXAMPLE 2—

A farmer wishes to make a small rectangular garden with one side against the barn. If the farmer has 200 feet of fencing, find the expression for the area of the garden.

First we make what I used to call my crummy little diagram before a real publisher got it. The picture has to be just good enough to see what is going on. A ruler is advisable.

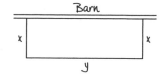

In the problem, we are told that we are looking for the area. We draw the picture and label the sides. The trick, if you can call it a trick, is that the fencing does not include the side of the barn.

So $2x + y = 200$. So $y = 200 - 2x$. The area $A = xy$. But $y = 200 - 2x$. The expression for the area in terms of x is $A = x(200 - 2x)$ or $200x - 2x^2$.

So far so good. Let's try another one, one not quite so nice.

EXAMPLE 3—

A potato farmer has a rectangular plot of land of 800 square feet. The three equal regions are pictured here.

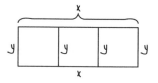

  A. Find an equation for the length of fencing.

  B. The outside fencing costs $20 a foot, while the inside is $7 a foot. Find an equation for the total cost.

  A. The area $A = 800 = xy$. The fencing $f = 2x + 4y$. From the area equation, $y = 800/x$. So $f = 2x + 4y$ can be rewritten $f = 2x + 4(800/x)$. $f = 2x + 3200x^{-1}$.

  B. The total cost is cost per foot times the number of feet. Outside fencing is $20 per foot. The number of feet is $2x + 2y = 2x + 2(800/x)$. Inside fencing is $7 per foot. The number of feet is $2y = 2(800/x)$. Total cost is $20[2x + 2(800/x)] + 7(2)(800/x) = 40x + 43,200x^{-1}$.

EXAMPLE 4—

An open box with a square bottom is to be cut from a piece of cardboard 10 feet by 10 feet by cutting out the corners and folding the sides up. Find the volume of the box.

We must cut squares out of the corners, dimension x by x. The pictures of the formation of the box are here:

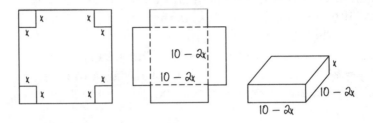

If the side is 10 and an x is cut off either end, the length and width are both $10 - 2x$. The height is x. $V = lwh = (10 - 2x)(10 - 2x)x$ or $4x^3 - 40x^2 + 100x$.

This one isn't too bad. The next one is a standard problem found in nearly all calc books.

EXAMPLE 5—

A box has a square base and no top.

    A. Find an expression for the surface area if the volume is 80.

    B. Find an expression for the volume if the surface area is 50.

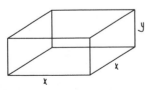

Let the base be of dimensions x by x and the height be y. The picture is as shown here.

    The volume is easy: $V = lwh = x^2y$. There are five sides for the surface area. The bottom is a square. Its area is x^2. The front, back, and two sides are all rectangles with the same dimensions x and y. The total surface area is $x^2 + 4xy$.

    A. $V = 80 = x^2y$. So $y = 80/x^2$.
       $SA = x^2 + 4xy = x^2 + 4x(80/x^2)$ or $x^2 + 320x^{-1}$.

B. SA $= 50 = x^2 + 4xy$. $y = (50 - x^2)/4x$.
  $V = x^2y = x^2(50 - x^2)/4x = (25x)/2 - x^3/4$.

EXAMPLE 6—

A 10-foot string is cut into two parts; one part is shaped into a circle and the other into a square. Find a formula that will give the sum of the areas.

Cut the string to make one piece x feet long. The remaining piece is $10 - x$ feet. The picture will look as shown here:

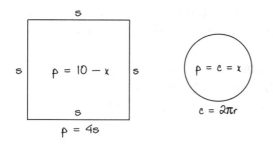

The circumference of the circle $c = x = 2\pi r$. So $r = x/2\pi$. Area of the circle $= \pi(x/2\pi)^2 = x^2/4\pi$.
  The perimeter of the square $p = 10 - x = 4s$. So $s = (10 - x)/4$.
  The area of the square is $s^2 = [(10 - x)/4]^2 = (x^2 - 20x + 100)/16$.
  The total area is $x^2/4\pi + (x^2 - 20x + 100)/16$.

Not too nice. Remember to keep plugging.
  Notice that Example 6 has one of the first answers with a π in the denominator.

EXAMPLE 7—

An orchard has 50 apple trees. The average number of apples per tree is 990. For each additional tree planted, the entire orchard gives 15 fewer apples per tree. Give an expression for the total number of apples.

The total number of apples as new trees are added equals the number of trees times the apples per tree. If we add x trees, the total number of trees is 50 + x. We lose 15 apples for each additional tree (or −15x). The number of apples per tree is 990 − 15x. The total number of apples is (50 + x) (990 − 15x) or 49,500 + 240x − 15x^2.

EXAMPLE 8—

A printer is using a broad page with 108 square inches. The margins are to be 1 inch on three sides and ½ inch at the top. Find an expression for the printed area.

Page area A = xy = 108; so y = 108/x. Length is x. With 2 inches cut off, the length of the print is x − 2. The width of the print is y − 1.5, 1½ inches cut off.

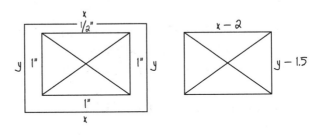

The area of the print is (x − 2)(y − 1.5) = (x − 2) (108x^{-1} − 1.5) or −1.5x + 111 + 216x^{-1}.

EXAMPLE 9—

Find the distance between the graph y^2 = 2x and the point (2, 0).

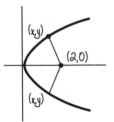

Distance means distance formula between the point (x, y) and (2, 0). d = [(x − 2)^2 + (y − 0)^2]$^{1/2}$ = (x^2 − 4x + 4 + y^2)$^{1/2}$.

But since $y^2 = 2x$, we can rewrite the distance formula $d = (x^2 - 4x + 4 + 2x)^{1/2}$ or $(x^2 - 2x + 4)^{1/2}$.

Here's another example that is found in most calc books.

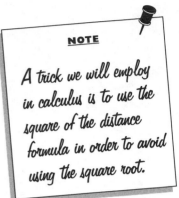

EXAMPLE 10—

A rectangle is inscribed in a parabola, $y = 12 - x^2$, with two vertices on the x axis and two vertices on the parabola. Find the expressions for the area and the perimeter of the rectangle in one variable, as usual.

When we draw the picture, the rectangle must be symmetrical with respect to the y axis.

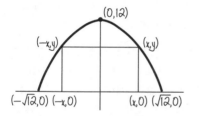

If (x, y) is any point on the curve, its symmetric point is $(-x, y)$. The base of the rectangle is $x - (-x) = 2x$, and the height is $y - 0 = y$.

The area $A = 2xy$. The perimeter is $4x + 2y$. However, since (x, y) is on the parabola, $y = 12 - x^2$. So $A = 2xy = 2x(12 - x^2)$ or $-2x^3 + 24x$:

$$p = 4x + 2y = 4x + 2(12 - x^2) \qquad \text{or} \qquad -2x^2 + 4x + 24$$

Hopefully these problems are now getting a little faster.

EXAMPLE 11—

A rectangle is to be inscribed in a 6-8-10 right triangle, so that the sides of the rectangle are parallel to the legs and one vertex lies on the hypotenuse. Find the equation of the area of this triangle in one variable.

There are three things we note about this problem:

1. We must set up the triangle in terms of the x-y axis, with the legs 6 and 8 on the x and y axes; it does not matter which leg is on which axis.

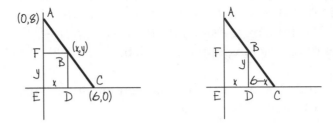

2. Wherever point B is, it represents a point (x, y) on the hypotenuse. The area of the rectangle is xy.

3. In order to find a relationship between x and y, we must see that triangles BCD and ACE are similar. (It is also possible to use triangle ABF.) Since $\overline{EC} = 6$ and $\overline{ED} = x$, $\overline{DC} = 6 - x$. The proportion we get is

$$\frac{\overline{BD}}{\overline{DC}} = \frac{\overline{AE}}{\overline{EC}} \quad \text{or} \quad \frac{y}{6-x} = \frac{8}{6}$$

Solving for y (by cross multiplication), we get

$$y = \frac{8(6-x)}{6} \text{ or } 8 - 4x/3$$

So $A = xy = x(8 - 4x/3)$ or $-4x^2/3 + 8x$.

We now start a number of rather nasty problems.

EXAMPLE 12—

Joan lives in an old house with a window that is in the shape of a rectangle surmounted with a semicircle.

If its perimeter is 10 meters, find the area of the window. The picture is as shown here.

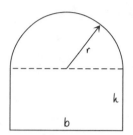

First, I don't know why they always say "surmounted," but they do. Next, there is a small trick to cut down on the fractions in this problem. We let b = 4x!!!!! We let h = y. Then r = ½b = 2x.

The perimeter is three sides of a rectangle (not the dotted side) plus a semicircle:

$$p = 2h + b + \frac{2\pi r}{2} = 2y + 4x + \pi(2x) = 10 \text{ so } y = 5 - 2x - \pi x$$

$$A = bh + \tfrac{1}{2}\pi r^2 = 4xy + \tfrac{1}{2}\pi(2x)^2 = 4x(5 - 2x - \pi x) + 2\pi x^2$$

$$= -2\pi x^2 - 8x^2 + 20x$$

They do not get any nicer. However, you should be exposed to these kinds of problems now, and most calculus professors do not give them.

EXAMPLE 13—

Write an equation for the volume of a cone inscribed in a sphere of radius 8. It must be in one variable, and the cone must touch the sphere in all possible places.

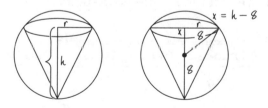

Again we note the symmetry of the cone inside the sphere. We see in the second picture here that we can find a relationship between r, x(= h − 8), and the radius of the sphere 8. Old Pythagoras tells us $(h - 8)^2 + r^2 = 8^2$. $r^2 = 16h - h^2$. The volume of a cone $V = (1/3)\pi r^2 h = (1/3)\pi(16h - h^2)h$.

The reason we substitute for r^2 and not for h is that the calculus problem associated with this problem is *much* easier this way.

EXAMPLE 14—

A man is on an island that is 4 miles from a straight shore. He wishes to go to a house that is 12 miles down the shore from the point that is closest to the island. Let x be the point where he lands on the shore. If he rows at 3 miles per hour and runs at 5 miles per hour, find the expression for the time it takes to get to the house.

The island I is 4 miles from the closest point C on the shore. We label as point P the point the man would row to. So $\overline{CP} = x$. Then $\overline{PH} = 12 - x$. IP is the Pythagorean theorem $(x^2 + 16)^{1/2}$.

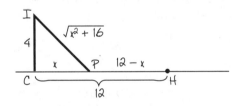

I consider this problem the ultimate rate × time = distance problem. We are looking for time = distance/rate.

	Distance	Rate	Time = Distance/Rate
From I to P (in the water)	$(x^2 + 16)^{1/2}$	3	$(x^2 + 16)^{1/2}/3$
From P to H (on land)	$12 - x$	5	$(12 - x)/5$

Total time $t = \dfrac{(x^2 + 16)^{1/2}}{3} + \dfrac{12}{5} - \dfrac{x}{5}$

The calculus part of this problem is not particularly nice either, although, as usual, it is the translation of the verbal problem into algebra that causes troubles.

There is one more problem we will do two ways: one using only algebra and one involving trig, as some of the problems in calculus do. The setting up, the part we will do here, should be studied. The calculus part is probably the nicest part, although it is not nice. The algebraic simplification is very messy. The trig simplification is really clever.

EXAMPLE 15—

A fence is 8 feet tall and is on level ground. The fence is parallel to a high building and is 1 foot from the building. A ladder over the fence touches the ground and the building. Find an expression for the length of the ladder.

NOTE

The picture is usually given. However, you do have to assign the variable x. The picture is as shown here.

ALGEBRAIC SOLUTION—

We are looking for the length of the ladder, $\overline{AE}$. We let $\overline{AB} = x$. From the problem, $\overline{BC} = 1$. By Pythagoras, $\overline{AD} = (x^2 + 64)^{1/2}$. We have similar triangles ABD and ACE.

$$\frac{\overline{AB}}{\overline{AD}} = \frac{\overline{AC}}{\overline{AE}} \qquad \text{or} \qquad \frac{x}{(x^2 + 64)^{1/2}} = \frac{x + 1}{\overline{AE}}$$

Using cross multiplication and then dividing by x, we get

Length of ladder $\overline{AE} = \dfrac{(x + 1)(x^2 + 64)^{1/2}}{x}$

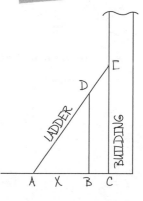

NOTE

If this seems like a strange problem, the calculus problem is to find the smallest ladder to do the trick.

TRIG SOLUTION—

We will do the problem in terms of the angle at A.

$\cot A = \overline{AB}/8 = x/8$ $x = 8 \cot A$

$\overline{AB} = x = 8 \cot A$ $\overline{AC} = \overline{AB} + 1 = 8 \cot A + 1$

$\sec A = \text{length of ladder}/\overline{AC} = \overline{AE}/\overline{AC}$

So $\overline{AC} \sec A = \text{length of ladder}$.

The length of the ladder is $(8 \cot A + 1) \sec A =$
$8 \csc A + \sec A$ since $(\cot A)(\sec A) = (\cos A/\sin A)$
$(1/\cos A) = 1/\sin A = \csc A$. (You must know those
identities.)

> **In writing trig functions, always try to get the unknown in the numerator. It makes calculations easier.**

If you work really hard on these problems, when you get to calculus you will be surprised how easy they will become.

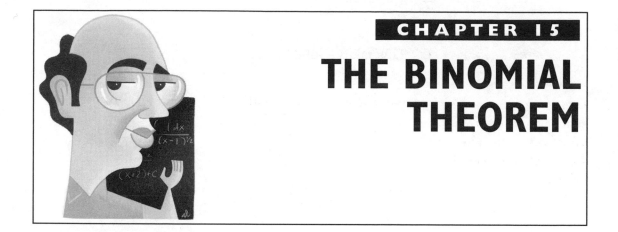

CHAPTER 15

THE BINOMIAL THEOREM

We would like to find the multiplied-out expression for, let's say, $(x + y)^8$ without saying "$(x + y)$ times $(x + y)$," getting the answer, multiplying again by $(x + y)$, and so on. This would take a very long time to do. There are a number of ways to find the expression. I think this method, looking at the pattern of each term, is the best:

$$(x + y)^n = x^n + \frac{n}{1} x^{n-1}y^1 + \frac{n(n-1)}{1(2)} x^{n-2}y^2$$

$$+ \frac{n(n-1)(n-2)}{1(2)(3)} x^{n-3}y^3$$

$$+ \frac{n(n-1)(n-2)(n-3)}{1(2)(3)(4)} x^{n-4}y^4 + \cdots$$

Do you see the pattern? If you do, you are pretty darn good. But if you don't see it or you're not sure you see it, a numerical example will help.

EXAMPLE 1—

Using the binomial theorem, write out and simplify all terms of $(2a - b)^5$.

This takes a lot of space:

$$(2a - b)^5 = (2a)^5 + \frac{5}{1}(2a)^4(-b)^1 + \frac{5(4)}{1(2)}(2a)^3(-b)^2$$

$$+ \frac{5(4)(3)}{1(2)(3)}(2a)^2(-b)^3 + \frac{5(4)(3)(2)}{1(2)(3)(4)}(2a)^1(-b)^4$$

$$+ \frac{5(4)(3)(2)(1)}{(1)(2)(3)(4)(5)}(2a)^0(-b)^5$$

We must make lots of observations:

1. If the power is n, and n is a positive integer, there are n + 1 terms. In Example 1 the power is 5; the number of terms is six.

2. If there is a minus sign, the *even* terms are negative (second, fourth, sixth, . . .).

3. Although we will not show it, it is possible for n to be a fraction or negative. In that case, we would get an infinite number of terms.

4. Notice the pattern of each term. For example, the fifth term in Example 1: The power of the second letter is 4, 1 less than the number of the term; 4 is the number of terms top and bottom in the coefficient; the bottom always starts at 1 and goes up to the power of the second term, in this case 4; the top starts at the power and goes down, and the number of terms is again the power of the second term, in this case 4 ($5 \times 4 \times 3 \times 2$); the powers of the letters add up to n, in this case 5—we have b^4, so a is to the first power; in the fifth term—odd number—the sign is plus.

Let's do another problem using the observations in item 4.

EXAMPLE 2—

Given $(3a - 5b)^{10}$. Write the fourth term only and simplify.

Well, we could write out all four terms, but that is a pain. Instead:

1. The sign is negative since it is the fourth term (even).

2. Since it is the fourth term, it contains $(5b)^3$.

3. Three is the number of terms on the bottom of the fraction: $1 \times 2 \times 3$.

4. Three on the bottom, three on the top starting with 10: $10 \times 9 \times 8$.

5. The power of $(5b)$ is 3. The total powers must add to 10. Soooooooo the power of $(3a)$ is 7.

The whole term is

$$-\frac{(10)(9)(8)}{(1)(2)(3)} (3a)^7(5b)^3 \quad \text{or} \quad -120(3a)^7(5b)^3$$

The coefficient of a^7b^3 is $120 \times 3^7 \times 5^3$. (Don't forget the coefficient part of each letter.) The whole answer is $-32,805,000a^7b^3$.

NOTE

The coefficient part of the term, in this case $10(9)(8)/(1)(2)(3)$, is always an integer. You should cancel first before you multiply.

ALSO NOTE

If you are asked for a term in which the first letter has the smaller exponent, use that number in writing the numerical coefficient.

One more example—a little trickier.

EXAMPLE 3—

$[x^2 + (1/x)]^6$. Write the term that contains no x.

We must write out two terms to see the pattern:

$$(x^2)^6 + \binom{6}{1}(x^2)^5\left(\frac{1}{x}\right)^1 + \cdots = x^{12} + 6x^9 + \cdots$$

The exponents go down by 3. So we are looking for the fifth term. It contains $(1/x)^4$. On the bottom of the fraction is $1 \times 2 \times 3 \times 4$. On the top of the fraction is $6 \times 5 \times 4 \times 3$ (four terms starting with 6 and going down). Since $(1/x)$ is to the fourth power and $4 + 2 = 6$, x^2 must be to the second power—$(x^2)^2$. So the term is

$$\frac{(6)(5)(4)(3)}{(1)(2)(3)(4)}(x^2)^2\left(\frac{1}{x}\right)^4 = 15x^4\left(\frac{1}{x^4}\right) = 15$$

Let's go on.

CHAPTER 16

THE THEORY OF EQUATIONS

The theory of equations is a topic that is increasingly found in precalculus courses. It consists of a series of theorems that will be listed and explained here but not proved.

A polynomial equation in x of the nth degree can be written in the form:

$$a_0 x^n + a_1 x^{n-1} + a_2 x^{n-2} + \cdots + a_n = 0$$

where $a_0 \neq 0$; $a_0, a_1, a_2, \ldots, a_n$ are numbers; and n is a positive integer.

A *root*, or *solution*, to this equation is a value of x that satisfies the equation. To *solve* an equation means to find all its roots.

Equations of degree 1, 2, 3, and 4 are called *linear*, *quadratic*, *cubic*, and *quartic*, respectively.

The coefficient a_0 is called the *leading coefficient* while a_n is the *constant term*.

THEOREM 1: THE FUNDAMENTAL THEOREM OF ALGEBRA

Every polynomial equation of degree n, n ≥ 1, n is a positive integer, has at least one root.

THEOREM 2: NUMBER OF ROOTS

Every nth degree polynomial equation has exactly n roots.

It appears that theorem 1 is very easy to prove while theorem 2 is very difficult. Nothing is further from the truth. In theorem 2, if there is one solution, factor it out. Do this n times, and the theorem is proved.

In this whole section, the only theorem that is hard to prove is the first one.

The fundamental theorem of algebra was first proved by the mathematician Gauss, and it can be proven relatively early after finishing the complete calculus sequence.

THEOREM 3: THE REMAINDER THEOREM

If $F(x)$, a polynomial equation of degree 1 or more, is divided by $x - r$, the remainder is $F(r)$.

THEOREM 4: THE FACTOR THEOREM

If $x - r$ is a factor of $F(x)$, then r is a root of the equation $F(x) = 0$ and conversely.

EXAMPLE I—

If $F(x) = x^3 + 2x^2 + 3x + 4$ is divided by $x - 2$, the remainder is $F(2) = 26$.

To check, long divide F(x) by x – 2 or use synthetic division. If you don't know long division, check *Algebra for the Clueless.* Don't know synthetic division? Check the appendix of this book.

EXAMPLE 2—

$F(x) = x^2 + 5x + 6 = 0$

$F(-2) = 0$. So $x - (-2) = x + 2$ is a factor of F(x).

EXAMPLE 3—

Show that 2 is a root of the equation $g(x) = 2x^3 + x^2 + -8x - 4 = 0$.

$$\begin{array}{r|rrrr} 2 & 2 & 1 & -8 & -4 \\ & & 4 & 10 & 4 \\ \hline & 2 & 5 & 2 & 0 \end{array}$$

Since f(2) is 0, x – 2 is a factor of f(x), and $f(x) = (x - 2)(2x^2 + 5x + 2)$. To find the rest of the roots, we can factor or use the quadratic formula.

The goal of all these theorems is to reduce the equation to a quadratic.

THEOREM 5: THE RATIONAL ROOT THEOREM

Given the polynomial equation $F(x) = a_o x^n + a_1 x^{n-1} + \cdots + a_n = 0$.

If (!!!) there is rational root p/q, p is a factor of a_n and q is a factor of a_n.

EXAMPLE 4—

Given the equation $F(x) = 4x^4 - 4x^3 - 13x^2 - 12x + 3 = 0$.

Possible numerators: ±1, ±3, factors of 3

Possible denominators: ±1, ±2, ±4, factors of 4

Possible rational roots: ±1, ±3, $\pm\frac{1}{2}$, $\pm\frac{3}{2}$, $\pm\frac{1}{4}$, $\pm\frac{3}{4}$

By trial and error, F(1/2) = 0. Synthetically dividing by 1/2 and dividing by 2, the remainder equation is $2x^3 - x^2 + 6x - 3 = 0$. Now the possible rational roots are ±1, ±3, $\pm\frac{1}{2}$, $\pm\frac{3}{2}$. Again F(1/2) = 0. Again using synthetic division, we get $2x^2 + 6$ or $x^2 + 3 = 0$. So the roots are ½, a double root, and $\pm i\sqrt{3}$.

EXAMPLE 5—

Prove $\sqrt{2}$ is irrational.

$\sqrt{2}$ satisfies the equation $f(x) = x^2 - 2 = 0$. The only possible rational roots are ±1, ±2 since the numerator is a factor of 2 and the denominator is a factor of 1; f(1), f(−1), f(2), and f(−2) are all not 0. Then f(x) is no rational root. But $\sqrt{2}$ is a root of this equation. $\sqrt{2}$ must be irrational! This is a really neat, short proof! Let's go on.

THEOREM 6: THE UPPER AND LOWER BOUNDS FOR ROOTS THEOREM

Given f(x) is a polynomial equation with leading coefficient positive.

If k, k > 0, has the final line all positive integers in synthetic division, then k is the upper bound for the roots. If we have the equation f(−x), and if k on the last line yields only positive coefficients, then −k is the lower bound for the roots.

EXAMPLE 6—

Given $g(x) = x^4 + x^3 + 70x^2 - 2x - 144 = 0$.

The possible rational roots are plus or minus 1, 2, 3, 4, 6, 8, 9, 12, 16, 18, 24, 36, 48, 72, and 144. Wow, horrible!! Using synthetic division, we find 2 yields all positive coefficients in the last line of synthetic division. Soooo 2 is the upper bound. For $g(-x)$, again we find 2 yields all last line positive coefficients. Soooo −2 is the lower bound. Since both 2 and −2 are not roots, if there is a rational root, it could be only 1 or −1. We'll get back to something like this later.

Let us look at Descartes' rule of signs.

THEOREM 7: DESCARTES' RULE OF SIGNS

Given $f(x) = 0$ has real coefficients:

1. The number of positive roots is not greater than the number of sign changes in $f(x) = 0$.

2. The number of negative roots is not greater than the number of sign changes in $f(-x) = 0$.

Corollary: If there are not exactly the number of roots (positive or negative) as the number of sign changes, it must be an even multiple less.

EXAMPLE 7—

Given $f(x) = x^4 - x^3 + 2x^2 - 7x + 5$ and $f(-x) = x^4 + x^3 + 2x^2 + 7x + 5$.

f(x) has 4 sign changes, so there is at most 4 positive roots.

If not 4, then 2. If not 2, then none.

f(–x) has no sign changes; so there are no negative roots.

The possibilities:

4 positive roots

2 positive roots and 2 complex (imaginary) roots

0 positive roots and 4 complex roots

Let's do one problem with all the tests.

EXAMPLE 8—

Find all the roots of $f(x) = 2x^4 + x^3 + 5x^2 + 4x - 12 = 0$

Theorem 1: According to the fundamental theorem of algebra, f(x) has at least one root.

Theorem 2: f(x) has exactly 4 roots.

Theorem 5: The possible rational roots are ±1, 2, 3, 4, 6, 12, ½, and ³⁄₂ since the numerators are factors of 12 and the denominators are factors of 2.

Theorem 6: By synthetic division of f(x) by x – 2, all coefficients in the last line are positive. The upper bound of the roots is 2. By synthetic division of f(–x) by x – 2, all coefficients in the last line are positive. The lower bound of the roots is –2. We need to look at only ±1, ±½, ±³⁄₂. *Note:* Always try the easiest roots first.

Theorem 4: f(1) = 0. Use synthetic division. We get $(x - 1)(2x^3 + 3x^2 + 8x + 12) = 0$. For $g(x) = 2x^3 + 3x^2 + 8x + 12 = 0$.

Theorem 7: There are no more positive roots and as many as 3 negative roots.

Theorem 7: There are 3 or 1 negative roots.

Theorem 3: f(−1/2) is not 0. It has a remainder. (x + 1/2) is not a factor.

Theorem 4: f(−3/2) = 0. Use synthetic division for −3/2. We get (x − 1)(2x + 3)(x^2 + 4) = 0. The roots are 1, −3/2, 2i, and −2i.

WHAT TO DO IF NOTHING ELSE WORKS

Finally, suppose we have a polynomial that does not have rational roots or easy-to-get irrational roots. Suppose we have a polynomial that has f(2) > 0 and f(3) < 0. Because f(x) is continuous (no breaks), there must be a root between 2 and 3. By trial and error, f(2.4) > 0 and f(2.5) < 0, there must be a root between 2.4 and 2.5. f(2.47) > 0, and f(2.48) < 0, which must mean that there is a root between 2.47 and 2.48. Continue the process until you get the accuracy you require. I have said that I don't like calculators, but in this case, thank G-d we have them. To do this by hand would be awful.

I think you've gotten the idea of the theory of equations. Let's go on to something different.

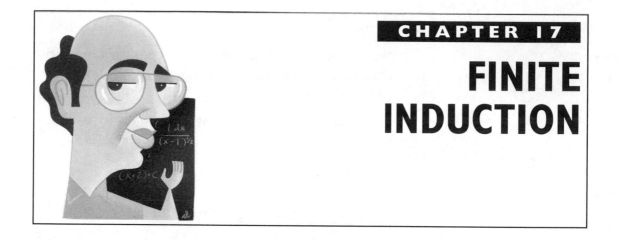

CHAPTER 17

FINITE INDUCTION

This subject used to be part of calculus and some pre-calculus courses. For completeness, it is a topic that should be covered. We will cover it here using two kinds of problems, which should be enough for basic calculus.

DEFINITION

The principle of finite induction: We prove a statement for $n = 1$. We assume a statement is true for $n = k$. We then prove the statement for $n = k + 1$. Then the statement is true for all positive numbers by finite induction.

Here is an example that may make the definition clear.

Imagine an infinite number of dominos in a line:

"Prove for $n = 1$" means if you knock over the first domino, it will knock over the next one.

"Assume for $n = k$" means that for some domino later, it will knock over the next one.

"Prove for n = k + 1" means that whenever the next one is knocked over, the one after will be knocked over. This means they all will fall down.

There are three formulas that are commonly used in calculus:

A. $1 + 2 + 3 + \cdots + n = \dfrac{n(n+1)}{2}$

B. $1^2 + 2^2 + 3^2 + \cdots + n^2 = \dfrac{n(n+1)(2n+1)}{6}$

C. $1^3 + 2^3 + 3^3 + \cdots + n^3 = \dfrac{n^2(n+1)^2}{4}$

The most famous formula is the first formula, which gives you the answer to adding the first n positive integers. It is attributed to Carl Friedrich Gauss (1777–1855), a very famous mathematician and physicist.

I have heard several versions of the story about this formula. My favorite is that Gauss discovered it at the age of $4\frac{1}{2}$ when his kindergarten class was noisy. The class was given the punishment of adding the first 100 positive integers. He saw in the sum $1 + 2 + 3 + \cdots + 98 + 99 + 100$ that $1 + 100 = 101$, $2 + 99 = 101$, and $3 + 98 = 101$ and that there were 50 such pairs. Soooo $50(101) = (100/2)(1 + 100) = 5050$!!!! Gauss discovered this at $4\frac{1}{2}$!!!!!!!! We will prove the third one.

EXAMPLE 1—

C: $1^3 + 2^3 + 3^3 + \cdots + n^3 = \dfrac{n^2(n+1)^2}{4}$

Prove for n = 1:

$$1^3 = \frac{1^2(1+1)^2}{4} \qquad 1 = 1$$

We have proven for $n = 1$.

Over 95% of the time, $n = 1$ is EZ. For clarity we will show $n = 5$.

For $n = 5$,

$$1^3 + 2^3 + 3^3 + 4^3 + 5^3 = \frac{5^2(5+1)^2}{4} \quad .225 = 225$$

Assume for $n = k$:

$$1^3 + 2^3 + 3^3 + \cdots + k^3 = \frac{k^2(k+1)^2}{4}$$

Prove for $n = k + 1$:

$$1^3 + 2^3 + 3^3 + \cdots + (k+1)^2 = \frac{(k+1)^2(k+2)^2}{4}$$

The proof for most sum problems is to take the assumption step and add the next term to both sides. *Proof:*

$$1^3 + 2^3 + 3^3 + \cdots + k^3 = \frac{k^2(k+1)^2}{4}$$

By assumption.

$$1^3 + 2^3 + 3^3 + \cdots + k^3 + (k+1)^3 = \frac{k^2(k+1)^2}{4} + (k+1)^3$$

Add the next term $(k + 1)^3$ to both sides.

$$1^3 + 2^3 + 3^k + \cdots + (k+1)^3 = \frac{k^2(k+1)^2}{4} + \frac{4(k+1)^3}{4}$$

On the left side the k^3 becomes part of the "dots."

$$= \frac{(k+1)^2(k^2+4(k+1))}{4}$$

A series of algebraic steps.

$$= \frac{(k+1)^2(k^2+4k+4)}{4}$$

So $1^3 + 2^3 + 3^3 + \cdots + (k+1)^3 = \frac{(k+1)^2(k+2)^2}{4}$

The statement is true for $n = k + 1$.

We've proved the statement for $n = 1$; we've assumed

the statement for $n = k$; and we've proved the statement for $n = k + 1$. Therefore the statement is true for all positive integers by finite induction.

Let's try one more.

EXAMPLE 2—

$6^n - 1$ is divisible by 5 for all positive integers n.

Prove for $n = 1$: $6^1 - 1 (= 6 - 1 = 5)$ is divisible by 5. I told you $n = 1$ is usually verrry easy.

Assume for $n = k$: $6^k - 1$ is divisible by 5.

Prove for $n = k + 1$: $6^{k+1} - 1$ is divisible by 5.

Proof:

$6^{k+1} - 1 = 6^{k+1} - 6^k + 6^k - 1$

Adding 0, $-6^k + 6^k$, is a common mathematical technique.

Factor 6^k out of the first two terms.

$$= 6^k(6 - 1) + 6^k - 1$$

$$= 6^k(5) + 6^k - 1$$

Now $6^k(5)$ is divisible by 5, and by assumption, $6^k - 1$ is divisible by 5. Therefore, its sum $6^{k+1} - 1$ is divisible by 5.

We've proved for $n = 1$, assumed for $n = k$, and proved for $n = k + 1$. $6^n - 1$ is divisible by 5 for all positive integers by induction.

Oh, let's do one more.

EXAMPLE 3—

$2^n > n$ for all positive integers.

Prove for $n = 1$: $2^1 > 1$; end step one.

Assume for $n = k$: $2^k > k$.

Prove for $n = k + 1$: $2^{k+1} > k + 1$.

Proof:

$2^k > k$; $2(2^k) > 2k$; $2^{k+1} > 2k = k + k > k + 1$ since $k > 1$ whenever k is bigger than 1. Sooo $2^{k+1} > k + 1$. Therefore $2^n > n$ is true for all positive integers.

Let's try something else.

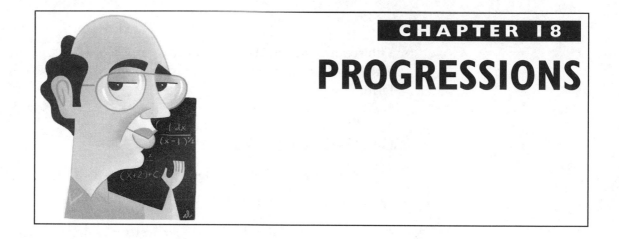

CHAPTER 18

PROGRESSIONS

In the twenty-first century, the SAT and the PSAT added progressions to their list of topics. Since progressions were always indirectly on these tests and they were always part of precalculus, this topic has been added to this book.

ARITHMETIC PROGRESSIONS

EXAMPLE 1—

4, 7, 10, 13, 16, . . . is an arithmetic progression because the difference between terms is the same, 3.

EXAMPLE 2—

100, 95, 90, 85, 80, . . . is an arithmetic progression because the difference between terms is the same, −5.

We need some notations:

The letter a stands for the first term.

The letter d stands for the common difference.

The second term is a + d. The third term is a + 2d. The fourth term is a + 3d. The last term, the nth term, *l* is a + (n − 1) d.

S denotes the sum.

Let's find the sum. It's a cute proof:

S = a + a + d + a + 2d + a + 3d + · · · + a + (n − 1) d

Starting from the other end:

S = *l* + *l* − d + *l* − 2d + *l* − 3d + · · · + *l* − (n − 1) d

Adding, we see allll the d's cancel. Since there are n terms and there is a + *l* in each term, we get 2S = n (a + *l*). The sum is

$$S = \frac{n}{2}(a + l)$$

Since *l* = a + (n − 1) d, substituting into S, we get a second expression for S:

$$S = \frac{n}{2}[2a + (n-1)d]$$

EXAMPLE 3—

Suppose the first term is 7 and the difference is 4. Find the 100th term and the sum.

a = 7, d = 5, n = 100. *l* = a + (n − 1) d = 7 + (100 − 1) 5 = 7 + 495 = 502.

$$S = \frac{n}{2}(a + l) = \frac{10}{2}(7 + 509) = 50(509) = 25{,}450$$

Let's try another.

EXAMPLE 4—

If the fifth term is 20 and the eighth term is 11, find the tenth term.

The eighth term is a + 7d = 11. The fifth term is
a + 4d = 20. Subtracting, we get 3d = − 9; d = − 3.

Substituting, say, into the fifth term, a + 4d = 20
becomes a + 4 (−3) = 20; a = 32. The tenth term is
a + 9d = 32 + 9 (−3) = 5. Let's try a couple more.

EXAMPLE 5—

In an arithmetic progression, if the first term is 1, the
difference is $\frac{5}{2}$, and the sum is $\frac{119}{2}$, find the number
of terms.

We use the formula

$$S = \frac{n}{2}[2a + (n - 1)d]$$

Then $\dfrac{119}{2} = \dfrac{n}{2}[2 + (n - 1)\tfrac{5}{2}]$

Multiplying each term by 4, we get 238 = n[4 + (n − 1)5]
or $5n^2 - n - 238 = 0$. Factoring, we get (5n + 34)
(n − 7) = 0. n can't be $-\frac{34}{5}$ since n is a positive integer;
therefore n = 7.

Some problems require a good knowledge of basic
algebra.

EXAMPLE 6—

a, $2a^2$, $3a^3$ is an arithmetic progression. Find a.

With three consecutive terms of an arithmetic progression, the middle term is the average (*arithmetic mean*)
of the other two terms. So

$$\frac{a + 3a^3}{2} = 2a^2$$

Simplifying we get $3a^3 - 4a^2 + a = 0$. Factoring, we get
a(3a − 1)(a − 1) = 0; a = 0, 1, $\frac{1}{3}$. If a = 0, we get the trivial
arithmetic progression 0, 0, 0. If a = 1, we get the

progression 1, 2, 3. If $a = \frac{1}{3}$, we get the progression $\frac{1}{3}$ $(= \frac{3}{9})$, $\frac{2}{9}$, $\frac{3}{27}$ $(= \frac{1}{9})$.

Let's try one for fun? Teachers have been known to make this a bonus on the progression test.

EXAMPLE 7—

Find a formula for the following sequence: 2, 6, 12, 20, 30, 42, ...

Look at the differences: $6 - 2 = 4$, $12 - 6 = 6$, $20 - 12 = 8$, $30 - 20 = 10$, $42 - 30 = 12$. The sequence of differences (4, 6, 8, 10, 12) forms an arithmetic progression. The original sequence is a quadratic function, $f(n) = an^2 + bn + c$. We need to find a, b, and c.

The third term is denoted by $f(3) = a(3)^2 + b(3) + c = 9a + 3b + c = 12$.

Similarly, the second term is $f(2) = a(2)^2 + b(2) + c = 4a + 2b + c = 6$.

Then the first term becomes $f(1) = a(1)^2 + b(1) + c = a + b + c = 2$.

Subtracting

$$f(3) = a(3)^2 + b(3) + c = 9a + 3b + c = 12$$

$$f(2) = a(2)^2 + b(2) + c = 4a + 2b + c = 6$$

we get

$$5a + b = 6$$

And subtracting

$$f(2) = a(2)^2 + b(2) + c = 4a + 2b + c = 6$$

$$f(1) = a(1)^2 + b(1) + c = a + b + c = 2$$

we get

$$3a + b = 4$$

Subtracting again, we get $2a = 2$. So $a = 1$. Substituting in, say, $3a + b = 4$, $b = 1$. Substituting $a = 1$ and $b = 1$ into $a + b + c = 2$, we get $c = 0$. So the formula is $f(n) = 1n^2 + 1n$ or just $n^2 + n$. Check out the remaining terms!

NOTE I

It is easier to do the third term first and put it on top when the terms increase.

NOTE 2

Allllways substitute into the easiest equation. In math, the simpler the better.

FINITE GEOMETRIC PROGRESSIONS

Over the years, this topic has become more important in precalculus. With the changes in the 2005 SAT, this topic becomes even more important since this is a different way of looking at exponential growth and decay problems. I like the topic for a different reason. This topic leads to infinite geometric progression, which is the last logical precalculus topic before calculus. Think about it. In the last section of this book, you are ready and able to start calculus! Let's begin.

The way you recognize a geometric progression is that the ratio between consecutive terms is the same.

EXAMPLE 8—

7, 70, 700, 7000, . . . is a geometric series since the ratio between terms is 10.

EXAMPLE 9—

100, 50, 25, 12½, . . . is a geometric series with ratio ½.

EXAMPLE 10—

7, −14, 28, −56, . . . is a geometric series with ratio −2.

Again the letter a is the first term; S is the sum; n is the number of terms; r is the ratio between terms; l is the last term.

The first term is a; the second term is ar; the third term is ar^2; and the last term $l = ar^{n-1}$.

Again, let us find expressions for S; the proof is nice:

Multiply each term by r and subtract.

$$S = a + ar + ar^2 + ar^3 + \cdots + ar^{n-1}$$
$$rS = ar + ar^2 + ar^3 + \cdots + ar^{n-1} + ar^n$$

All the terms in the middle cancel. What is left is

$$S - rS = a - ar^n. \quad S(1 - r) = a(1 - r^n).$$

$$S = \frac{a(1 - r^n)}{1 - r}$$

Since ar^n can also be written as rl, the sum also is

$$S = \frac{a - rl}{1 - r}$$

EXAMPLE 11—

In a geometric progression, a = 2, l = 486, and n = 6. Find r and S.

$$l = ar^{n-1}. \quad 486 = 2 \, (\, r^{6-1}). \quad r^5 = 243. \quad r = 3.$$

$$S = \frac{a - rl}{1 - r} = \frac{2 - 3(486)}{1 - 3} = 728$$

EXAMPLE 12—

Suppose the sixth term is 6400 and the third term is 800; find the eighth term.

The sixth term is $ar^5 = 6400$. The third term is $ar^2 = 800$. Dividing both sides, we get $r^3 = 8$. $r = 2$. Substituting into $ar^2 = a(2)^2 = 800$. $a = 200$. The eighth term $= ar^7 = (200)(2^7) = 25,600$.

EXAMPLE 13—

Find the geometric mean of 3 and 5.

The *geometric mean* means we have a geometric progression 3, x, 5, where x is the geometric mean.

   The ratio $r = x/3 = 5/x$. $x^2 = 15$. $x = \pm\sqrt{15}$.

INFINITE GEOMETRIC PROGRESSIONS

In the finite geometric progression, the numbers get really big or really small quickly. There are usually not too many problems. This is not true for the infinite geometric progression. However, we cannot always have an infinite geometric progression. Let's see when we can.

EXAMPLE 14—

Suppose $r > 1$, say, 2. For example, $6 + 12 + 24 + 48 + 96 + \cdots$.

Clearly the sum would be "infinite." So r can't be greater than 1.

EXAMPLE 15—

Suppose $r < -1$, say, -3. For example, $10 - 30 + 90 - 270 + 810 - 243 + \cdots$.

The sum would wildly alternate between a big answer and a small (very negative) answer. So r can't be less than -1.

EXAMPLE 16—

Suppose r = 1. For example, $5 + 5 + 5 + 5 + 5 + \cdots$.

Again the sum is "infinite."

EXAMPLE 17—

Suppose r = $-$ 1. For example, $5 - 5 + 5 - 5 + 5 - 5 + \cdots$.

The sum is either 5 or 0, depending on where you stop. So r can't be greater or equal to 1 or less than or equal to $-$ 1.

EXAMPLE 18—

Suppose $-1 < r < 1$, say, $\frac{1}{2}$. What is $(1/2)^n$?

As n gets very big, $(\frac{1}{2})^n$ gets verry small. Oh, let's see it: $\frac{1}{2}, \frac{1}{4}, \frac{1}{8}, \frac{1}{16}, \ldots$. Its limit is 0. $(\frac{1}{2})^n$ goes to zero.
    The expression

$$S = \frac{a(1-r^n)}{1-r}$$

becomes

$$S = \frac{a}{1-r}$$

the infinite geometric sum.

NOTE 1

We will explore the term *limit* in Chapter 19.

NOTE 2

There is not an infinite arithmetic sum since the sum would always go to "$\pm$ infinity."

Let's try some examples.

EXAMPLE 19—

r = ¾ and a = 10. Find S.

$$S = \frac{a}{1-r} = \frac{10}{1-3/4} = 40$$

EXAMPLE 20—

We all know how to change 1/3 to a decimal. We divide 3 into 1.0000:

$$3\overline{)1.0000} = .\overline{3} \qquad .3333$$

Let's change $.\overline{3}$ back to 1/3. You may never have done this:

$$.3333 = .3 + .03 + .003 + .0003 + \cdots$$

So a = .3 and r = .1:

$$S = \frac{a}{1-r} = \frac{.3}{1-.1} = \frac{.3}{.9} = \frac{3}{9} = \frac{1}{3} \;\;!!!!$$

EXAMPLE 21—

Change .787878 . . . to a fraction.

$$.787878 \ldots = .78 + .0078 + .000078 + \cdots$$

So a = .78 and r = .01, since it repeats every two places.

EXAMPLE 22—

Change .50123123123 . . . to a fraction.

In this problem, there is a nonrepeating part and a repeating part that must be separated:

$$.50123123123 \ldots = .50 + .00123 + .00000123 + \cdots$$

$.50 = \frac{1}{2}$. For the rest, $a = .00123$ and $r = .001$ since the repeating part repeats every three places:

$$.50123123123\ldots = \frac{1}{2} + \frac{a}{1-r} = \frac{1}{2} + \frac{.00123}{1-.001} = \frac{1}{2} + \frac{.00123}{.999}$$

$$= \frac{1}{2} + \frac{123}{99,900} = \frac{49,500}{99,900} + \frac{123}{99,900}$$

$$= \frac{49,623}{99,900} = \frac{16,541}{33,300}$$

EXAMPLE 23—

A ball is dropped from a height of 300 feet and bounces up $\frac{2}{3}$ of its previous height. If the ball does this forever, find the (up and down) distance it travels.

If you draw the picture, you will see two geometric sums.
   First, 300 feet is the first distance.
   Then you have two geometric sums, up and down.
   With the first term in each, $(\frac{2}{3})\, 300 = 200$, and the ratio is $\frac{2}{3}$. We get

$$300 + 2\left(\frac{a}{1-r}\right) = 300 + 2\left(\frac{200}{1-2/3}\right) = 300 + 1200 = 1500 \text{ feet}$$

We are now ready for limits.

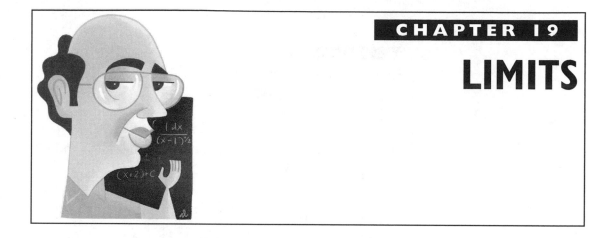

CHAPTER 19

LIMITS

Congratulations a second time. The first time was for reaching precalculus, a level that the vast majority of students never reach. With limits, you are starting calculus!!!!!

Actually we have already talked about limits. The sections on infinite series and asymptotes indirectly talked about limits. In this section we will be a little more formal.

FINITE LIMITS

We say the *limit of f(x) as x goes to a is L,* which we write as

$$\lim_{x \to a} f(x) = L$$

If the closer x gets to a, the closer f(x) gets to L.

The first two examples and half of Example 4 are given for understanding. The second half of Example 4 and beyond are examples of how you will do the problems.

EXAMPLE 1—

Find

$$\lim_{x \to 5} 4x$$

There are two ways to approach 5: numbers less than 5 and greater than 5.

A. x	4x	B. x	4x
4	16	6	24
4.5	18	5.5	22
4.9	19.6	5.1	20.4
4.99	19.96	5.0001	20.0004

In chart A, we see that the closer we get to 5 from the left, the closer the values get to 20. We write

$$\lim_{x \to 5^-} 4x = 20$$

In chart B, we see the closer we get to 5 from the right, the closer the values get to 20. We write

$$\lim_{x \to 5^+} 4x = 20$$

The limit from the left equals the limit from the right:

$$\lim_{x \to 5^-} 4x = \lim_{x \to 5^+} 4x$$

We say the limit exists and equals 20, and we write

$$\lim_{x \to 5} 4x = 20$$

After seeing this, you might say, "You big dummy! All you have to do is stick in 5 and get 20. Why did you go through so much work?!" My answer is, "If calculus were that easy, either you or I would have been able to invent it." I don't know about you, but I wouldn't have been able to do that.

Let's try another example.

EXAMPLE 2—

$$\lim_{x \to 8} \frac{x-8}{x-8}$$

Let us put in 8. We see we get 0/0. So substituting doesn't always work.

Let us again make charts.

A.

x	$\frac{x-8}{x-8}$
8.1	1
8.01	1

B.

x	$\frac{x-8}{x-8}$
7.9	1
7.99	1

We see that whether we take numbers to the left of 8 or the right of 8, the value is 1. Sooooo

$$\lim_{x \to 8} \frac{x-8}{x-8} = 1$$

even though

$$\frac{x-8}{x-8}$$

is not defined when x = 8.

The word *limit* means very, very close to the number but not necessarily at the number. This is a very difficult concept and must be looked at many times. If we don't get 0/0, the limit is what it is.

EXAMPLE 3—

A. $\lim_{x \to 5} \dfrac{2x+3}{3x+2} = \dfrac{13}{17}$

B. $\lim_{x \to 6} \dfrac{x-6}{x+1} = \dfrac{0}{7} = 0$

C. $\lim_{x \to 3} \dfrac{x-5}{x-3} = \dfrac{-2}{0}$ undefined

NOTE

The statement, "If we don't get 0/0, the limit is what it is" is not quite true. In higher math, there are complicated functions for which this is not true.

Also, if a function is defined in pieces, as done earlier in this book, substituting might not be the limit.

Let us do one last example for understanding and then illustrate how you will do the problems.

EXAMPLE 4—

$$\lim_{x \to 5} \frac{x^2 - 25}{5x - 25}$$

If we let $x = 5$, we get 0/0. For the last time we will make charts:

A.

x	$\dfrac{x^2 - 25}{5x - 25}$
6	2.2
5.5	2.1
5.1	2.02
5.01	2.002

B.

x	$\dfrac{x^2 - 25}{5x - 25}$
4	1.8
4.5	1.9
4.9	1.98
4.99	1.998

Two things seem clear: The limit is 2, and we never want to make tables.

EXAMPLE 4 REVISITED—

$$\lim_{x \to 5} \frac{x^2 - 25}{5x - 25}$$

We must do something algebraic. In this case we factor:

$$\lim_{x \to 5} \frac{x^2 - 25}{5x - 25} = \lim_{x \to 5} \frac{(x + 5)(x - 5)}{5 \ (x - 5)}$$

By Example 2,

$$\lim_{x \to 5} \frac{x-5}{x-5} = 1$$

Also as x goes to 5, x + 5 goes to 10. Sooo

$$\lim_{x \to 5} \frac{(x+5)(x-5)}{5 \quad (x-5)} = \frac{10}{5}(1) = 2$$

EXAMPLE 5—

$$\lim_{x \to 4} \frac{\frac{1}{x} - \frac{1}{4}}{x^2 - 16}$$

In this problem we multiply the numerator and denominator by 4x and factor.

$$\frac{\frac{1}{x} - \frac{1}{4}}{x^2 - 16} = \frac{\left(\frac{1}{x} - \frac{1}{4}\right)}{(x-4)(x+4)} \left(\frac{4x}{4x}\right) = \frac{4-x}{(x-4)(x+4)4x}$$

Now

$$\lim_{x \to 4} \frac{4-x}{x-4} = -1$$

since $(4-x) = -1(x-4)$. Also, as x gets close to 4, x + 4 gets close to 8 and 4x gets close to 16. So

$$\lim_{x \to 4} \frac{4-x}{(x-4)(x+4)4x} = \frac{-1}{8(16)} = \frac{-1}{128}$$

EXAMPLE 6—

Let

$$f(x) = \begin{array}{ll} 2x & \text{if } x < 2 \\ ax + b & \text{if } 2 \le x \le 4 \\ x^2 & \text{if } x > 4 \end{array}$$

Find a and b so that the limit exists at x = 2 and x = 4.

We use the fact that the limit from the left must equal the limit from the right:

$$\lim_{x \to 2^-} 2x = 4 \text{ and } \lim_{x \to 2^+} (ax + b) = 2a + b \qquad \text{So } 2a + b = 4.$$

$$\lim_{x \to 4^-} (ax + b) = 4a + b \text{ and}$$

$$\lim_{x \to 4^+} x^2 = 16 \qquad \text{So } 4a + b = 16.$$

$$4a + b = 16$$
$$2a + b = 4$$

Subtracting, we get $2a = 12$. $a = 6$. Substituting, we get $b = -8$.

LIMITS GOING TO INFINITY

The limit of $f(x)$ as x goes to infinity, is written

$$\lim_{x \to \infty} f(x) = L$$

if the bigger x gets, the closer $f(x)$ gets to L.

Again, we talked about this when we talked about asymptotes. Let's talk a little more.

EXAMPLE 7—
Find

$$\lim_{x \to \infty} \frac{3x^2 + 4x + 5}{5 - 9x^3}$$

If you recall earlier in the book, this is exactly the way we found Horizontal asymptotes. Since the degree of the top is less than the degree of the bottom, the limit is 0!

EXAMPLE 8—
Find

$$\lim_{x \to \infty} \frac{5x^2 - 7x}{4 - 9x^2}$$

Again, we did this before. Since the degree of the top equals the degree of the bottom, the limit is the coefficient of the highest power on the top divided by the coefficient of the highest power on the bottom. In this case the limit is 5/−9 or −5/9. OK, let's do a new problem!!!!

EXAMPLE 9—

Find

A. $\displaystyle \lim_{x \to \infty} \frac{\sqrt{5x^2 + 6x + 7}}{3x}$

B. $\displaystyle \lim_{x \to -\infty} \frac{\sqrt{5x^2 + 6x + 7}}{3x}$

If we could talk about the degree of each numerator, it would be considered 1: x^2 to the ½ power. For very large values of x, say, x = 1,000,000, or for small values of x, say, x = −1,000,000, only the highest power need be counted since $(1{,}000{,}000)^2$ is muuuuch bigger than 1,000,000. So we can approximate $\sqrt{5x^2 + 6x + 7}$ as $\sqrt{5x^2} = \sqrt{5}|x|$. So Example 9A becomes

$$\lim_{x \to \infty} \frac{\sqrt{5}|x|}{3x} = \frac{\sqrt{5}}{3}$$

and 9B becomes

$$\lim_{x \to -\infty} \frac{\sqrt{5}|x|}{3x} = \frac{-\sqrt{5}}{3}$$

EXAMPLE 10—

Find

$$\lim_{x \to \infty} \left(\sqrt{x^2 + 9} - \sqrt{x^2 - 6} \right)$$

Here we get infinity minus infinity. We must do something. 99½% of the time, when we have two square roots subtracted or added, we think of conjugate, even though there is no denominator. So

$$\left(\frac{\left(\sqrt{x^2+9}-\sqrt{x^2-6}\right)}{1} \times \frac{\left(\sqrt{x^2+9}+\sqrt{x^2-6}\right)}{\left(\sqrt{x^2+9}+\sqrt{x^2-6}\right)}\right) = \frac{15}{\sqrt{x^2+9}+\sqrt{x^2-6}}$$

The degree of the top, 0, is less than the degree of the bottom (if it had a degree). The limit is zero.

This seems like a really strange place to end the book, but this is the beginning of Calc I. If you want to learn calculus, read my *Calc for the Clueless* series, *Calc I*, *Calc II*, and *Calc III*. For most of you my *SAT Math for the Clueless*, second edition, will be useful. Until then, have a nice summer!

SYNTHETIC DIVISION AND SETS

SYNTHETIC DIVISION

EXAMPLE 1—

Divide: $2x^3 - 4x + 5$ by $x + 2$.

All of the powers of x must be written highest to lowest with no powers missing.

We have $2x^3 + 0x^2 - 4x + 5$.

Next, we write the term that makes $x + 2 = 0$, $x = -2$. If we had $2x - 7 = 0$, we would make $x - 7/2 = 0$ and take $+7/2$.

Next put -2 in a box. Next to it write the coefficients:

$$
\begin{array}{r|rrrr}
-2 & 2 & 0 & -4 & 5 \\
   &   & -4 & 8 & -8 \\
\hline
   & 2 & -4 & 4 & -3 \\
\end{array}
$$

1. Bring down the 2.

2. $(-2)(2) = -4$.

3. $0 + (-4) = -4$.

4. $(-2)(-4) = 8$.

5. $(-4) + 8 = 4$.

6. $(-2)(4) = -8$.

7. $5 + (-8) = -3$.

The answer starts one power less than the original. It is $2x^2 - 4x + 4 + (-3)/(x + 2)$.

SETS

We have talked a little about sets with the range and domain. The twenty-first-century SAT requires more.

You cannot actually define a set. A *set* is a "collection" of things, called *elements*.

In the set {1, 2, 3}, we write $3 \in$ {1, 2, 3}, read "3 is an element in the set {1, 2, 3}."

We write $4 \notin$ {1, 2, 3}, read "4 is not an element in the set." Note that $\in$ is the Greek letter epsilon.

We write $A \cup B$, read "A union B is the set of elements in A or in B or in both."

We write $A \cap B$, read "A intersect(ion) B is the set of elements in both sets."

EXAMPLE 2—

Let A = {1, 2, 3, 4, 5}, B = {1, 3, 5, 7}, and C = {2, 4, 9}.

Then $A \cup B$ = {1, 2, 3, 4, 5, 7}. $A \cap B$ = {3, 5}.

$B \cap C$ = { }, the *null set*, the set with no elements. It is usually written Ø, which is the Greek letter phi, pronounced fee or fie.

NOTE

{0} is not, not, not the null set. It is the set with one element, namely the number 0.

Two sets are equal if they have exactly the same elements.

EXAMPLE 3—

$\{3, 4\} = \{4, 3\}$.

Order doesn't matter.

$\{4, 3\} = \{3, 3, 3, 3, 4, 3, 4, 4\}$.

Repeated elements don't count. Each set here has two elements.

A is a subset of B, which is written $A \subseteq B$, if every element in A is always in B. If a set has n elements, it has 2^n subsets.

EXAMPLE 4—

Write all the subsets of $\{a, b, c\}$.

There are 2^3, or 8, subsets. Let's write them:

$\emptyset$, $\{a\}$, $\{b\}$, $\{c\}$, $\{a, b\}$, $\{a, c\}$, $\{b, c\}$, $\{a, b, c\}$

In every discussion of sets, there is a *universe*, U, which includes everything under discussion.

Suppose $U = \{1, 2, 3, 4, 5, 6\}$. A^c, read "A complement," is all the elements in the universe not in A:

If $A = \{2, 6\}$, then $A^c = \{1, 3, 4, 5\}$.

NOTE 1

There are many notations for the complement of a set.

NOTE 2

Change the universe and you change the complement.

NOTE 3

This complement is spelled with an **e**. Compliment, how good looking and smart all you readers are, is spelled with an **i**, a compliment.

What a wonderful way to end the book! Have a great life!

INDEX

ABOUT BOB MILLER ... IN HIS OWN WORDS

I received my B.S. and M.S. in mathematics from Poly-technic University in New York after graduating from George W. Hewlett High School, Hewlett, New York. After my first class, which I taught as a substitute for a full professor at Poly, one student said to another upon leaving the room, "At least we have someone who can teach the stuff." I was forever hooked on teaching math.

Since then I have taught at the City University of New York (CUNY), Rutgers, and Westfield State College in Massachusetts. My name has been included in three editions of *Who's Who Among America's Teachers*.

No matter how badly I feel, I always feel better after I teach. I am always delighted when students tell me they hated math before but now they like it and can do it.

I have a fantastic wife Marlene, tremendous children Sheryl, Eric, Glenn, and Wanda, and unbelievable grandchildren, Kira, Evan, Sean, and Sarah. My hobbies are golf, bowling, bridge, and crossword puzzles.

To me, teaching math is a great joy. I hope I can give some of that joy to you.